EINSTEIN'S THIRD MISTAKE

PRINCIPLE OF EQUIVALENCE

EVGENI BANTUTOV

Copyright © 2023 EVGENIUS

All rights reserved

The characters and events portrayed in this book are fictitious. Any similarity to real persons, living or dead, is coincidental and not intended by the author.

No part of this book may be reproduced, or stored in a retrieval system, or transmitted in any form or by any means, electronic, mechanical, photocopying, recording, or otherwise, without express written permission of the publisher.

CONTENTS

Title Page
Copyright
1. Introduction. 1
2. Definitional area. 3
3. Principle of equivalence. 5
4. Newton's first law. 15
5 . Newton's second law. 24
6. Newton's third law. 34
7. Newton's Law of Gravitation. 46
8. Relative motion at constant velocity. 49
9. Absolute motion with constant acceleration. 53
10. Attribution of types of movements. 58
11. Sensation of the action of force. 82
12. Strength. Application point of action. 89
13. Types of forces. Manifestation of power. Cause effect. 90
14. Principle of uniformity. 95
15. Graphic representation 98
16. Condition of relative rest 103
17. Three-dimensional reality. One dimensional reality. 109
18. Effort. Acceleration. 124
19. Field of effort. Common fundamental essence of the 130

One Infinite Reality.
20. Newton, gravity and field of effort .   141
21 TIME   143

1. INTRODUCTION.

This book is written for readers who do not have a special education in Physics.

There are many figures that show and explain the problems of modern physics. There are no complicated mathematical formulas. It is shown that much of the problems of modern physics are caused by the Theory of Relativity, which was created by Einstein.

Einstein noticed that when a body is moving with acceleration in a gravitational field, its accelerating motion is identical to uniform rectilinear motion.

Einstein used this fact and then, motion with acceleration can be equated to uniform rectilinear motion. This means that the two types of motion are equivalent, and Einstein defined it as *the Principle of Equivalence*.

Einstein equated accelerating motion with uniform rectilinear motion, and thus created the General Theory of Relativity.

The opposite should be done. Uniform rectilinear motion must be equated with accelerating motion. Then, uniform rectilinear motion is equivalent to motion with acceleration. Then, uniform rectilinear motion is a special case of motion with acceleration.

Einstein defined the Principle of Equivalence, and created the General Theory of Relativity. The Principle of Equivalence is incorrectly defined. This creates huge problems for the Theory of Relativity, and a crisis in modern physics.

To create General Relativity, the Principle of Equality must be used.

It follows from the Principle of Equality that:

The force of gravitational attraction as defined by Newton **is not** a central force. Newton's force of gravitational attraction is a transversely acting force.

Newton's law of gravitation is true only within the confines of the solar system.

Then Dark Energy and Dark Matter do not exist.

There are an infinite number of different **"laws of gravity"**, and these laws are realized in **a field of effort**.

The field of effort is the carrier of the higher derivatives of distance and time.

The action *MUTUALISACTION* takes place in **the field of effort**.

Translation from Slavic - Bulgarian Cyrillic, to English:

| ВЗАИМНОДЕЙСТВИЕ = MUTUALISACTION |

2. DEFINITIONAL AREA.

An analysis of the basic laws of Physics will be carried out. To perform the analysis correctly, it is necessary to create a suitable definition area. The definitional domain consists of four axiomatic principles and one philosophical category.

Principles:

1- Reality **exists**.

2- Reality is **reflective**.

3- Reality is **infinite**.

4- Reality is single, unique.

Philosophical category:

The concept of **the One Infinite Reality** is a philosophical category.

Explanations:

- The concept of **One Infinite Reality** is a philosophical category that serves to denote the unity of consciousness and matter.

-**Existence** is an independent category of science philosophy. Non-philosophers usually antagonistically oppose the category of existence to the category of non-existence. It is usually answered that what does not exist is called nothing. The next step is to analyze the categories **nothing** and **something**. The analysis of these two categories is extremely difficult, and the conclusions are incorrect.

In the hypothesis I present, **existence** is not opposed to non-existence. Existence is an additional category to the **reflection**

category.

Existence and **Reflection** are a pair of categories.

In the hypothesis I present, existence and reflection have been added to the pairs of categories of Hegel's Dialectic.

See Hegel, Phenomenology of Spirit.

See Todor Pavlov, "Theory of Reflection".

- The category **Infinity** serves to indicate the infinite amount of existing qualities.

- The category **Single** serves to indicate the uniqueness of **the universal**.

The category **Single** is present in the system of Hegel's Dialectical Logic.

The category **Singular** is part of Hegel's three categories: **singular**, **special**, **general**. See Hegel, Phenomenology of Spirit.

3. PRINCIPLE OF EQUIVALENCE.

The Principle of Equivalence was defined by Albert Einstein. Einstein used the Equivalence Principle to create the General Theory of Relativity. The Principle of Equivalence states that:

-the heavy and inert mass of any physical body are equal and that:

- the motion of a body with acceleration in a gravitational field is equivalent to uniform rectilinear motion .

These are two important facts that are placed in the foundations of the General Theory of Relativity. I will use figures to explain these two facts. I begin by explaining the equality of heavy and inertial mass.

See Figure 1.

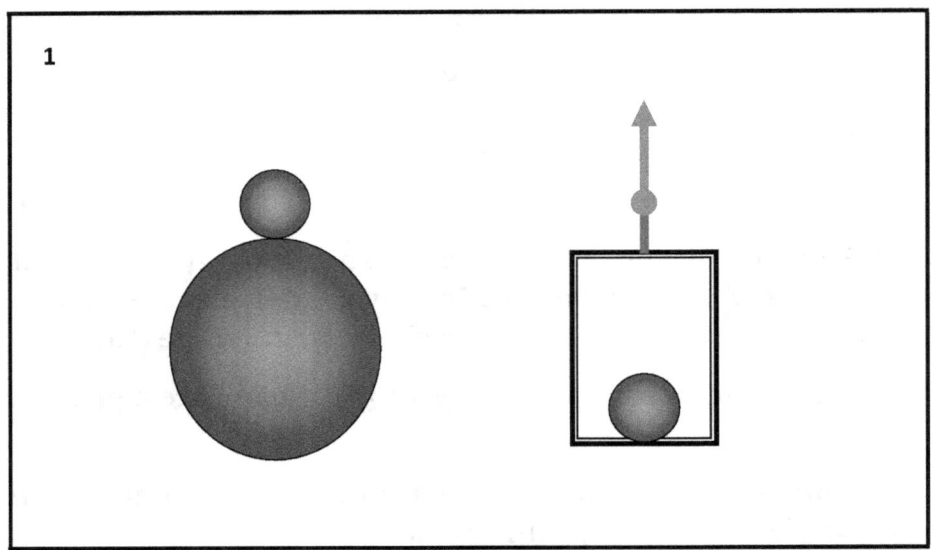

In the left part of figure 1, two spheres, small and large, are shown. The small sphere is placed on top of the large sphere. In the right part of figure one, an elevator is shown, and once again, the same small sphere that is placed at the bottom of the elevator.

The elevator and the small sphere are located in outer space, where no gravitational forces act.

The great sphere is the planet Earth. The small sphere is a test body that is located on the surface of the planet Earth. The small sphere has some weight which is called **heavy mass** . The small sphere that is on the surface of planet Earth is exactly the same as the small sphere that is placed at the bottom of the elevator. The elevator is attached to a brown rope. At the end of the brown rope, there is a red force acting that pulls the elevator in the direction shown. The force applied to the end of the rope is of such magnitude that the elevator moves with an acceleration equal to nine whole and eight tenths meters per second squared. When the elevator is moving in the direction shown with an acceleration equal to nine whole eight tenths of a meter per second squared, the small sphere at the bottom of the elevator will have weight.

This weight is called **inertial mass**.

The heavy mass of the small sphere that is located on the surface of the planet Earth is equal to the **inertial mass** of the small sphere that is located at the bottom of the elevator.

See Figure 2.

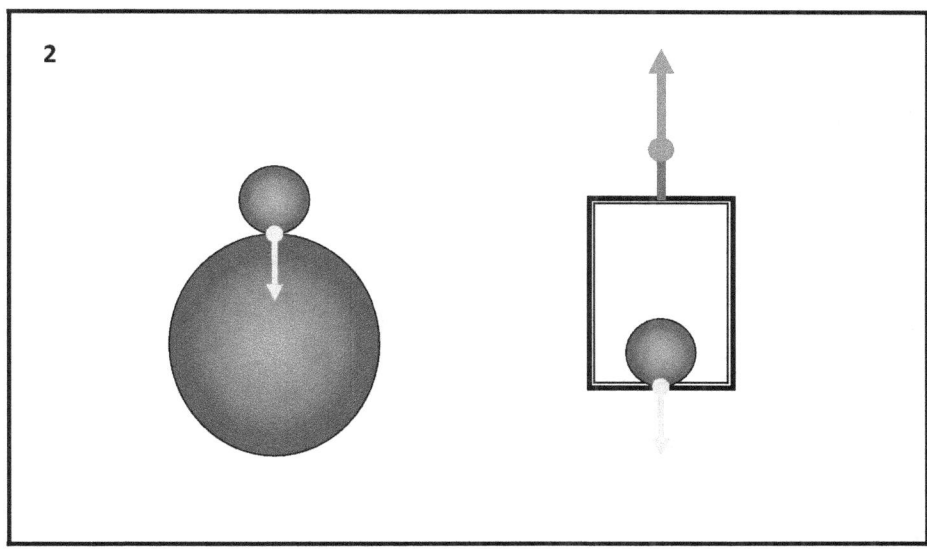

In Figure 2, the small sphere on the surface of the planet Earth is shown, pressing down on the Earth's surface by its **heavy mass**. The green arrow is the pressure force. Shown is the small sphere in the elevator pushing the bottom of the elevator through its **inertial mass**. The green arrow below the lift indicates the magnitude and direction of the push. The two small spheres are the same, the length of the green arrows is the same, which means that **the gravity and inertial mass** of the small sphere are the same.

The reason for the equality of **the heavy and inertial masses** is the fact that the earth's gravitational acceleration is equal to nine whole eight tenths of a meter, per second squared, and the acceleration with which the elevator moves in a vertical direction is also equal to nine whole eight tenths meters, per second per square.

In short, **heavy mass** is always equal to **inertial mass**.

We can verify the equality of heavy mass and inertial mass. We use two accurate scales.

See Figure 3.

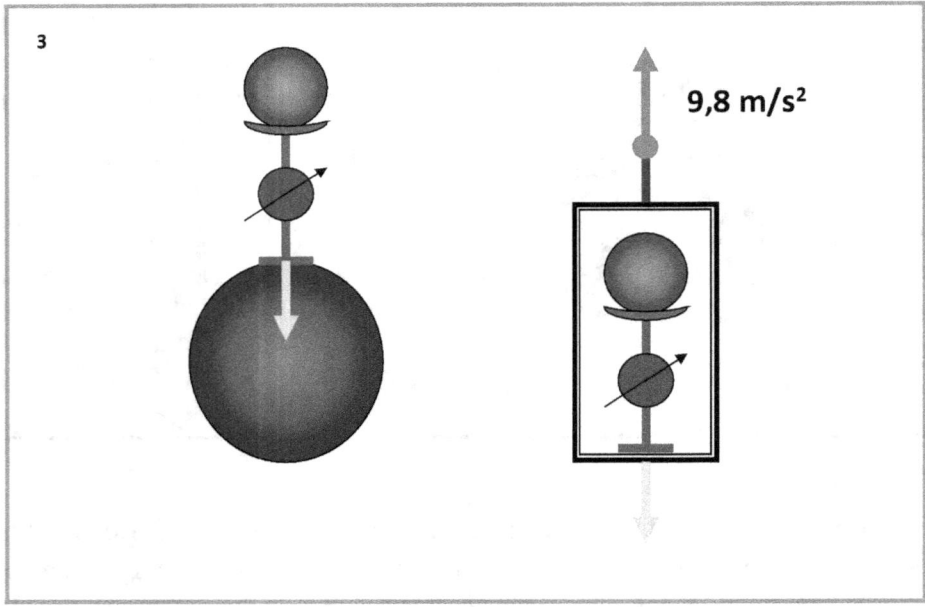

Figure 3 shows two identical scales. The scales have a blue display for weight reading, a brown base and a brown support plate.

Look at the left side of the picture. The base of the scale is on the earth's surface. Above the scale is placed the small sphere. The black arrow indicates the weight of the small sphere. A scale placed on the earth's surface measures **the heavy mass** of the small sphere.

The same scale is placed on the bottom of the elevator. The small sphere is placed on the scale. The black arrow indicates the weight of the small sphere. The scale in the elevator measures **the inertial mass** of the small sphere. Black arrows on both scales indicate equal weight. **The heavy mass** of the small sphere is equal to **the inertial mass** of the small sphere. The bases of both scales press down equally. The two green arrows below the bases of the scales are the same length.

The second important fact in the Principle of Equivalence is that:

- the motion of a body with acceleration in a gravitational field is equivalent to uniform rectilinear motion .

To explain this fact, we will conduct a thought experiment, with an elevator and a passenger who moves along with the elevator. Unfortunately, at some point in time, the rope breaks. See Figure 4.

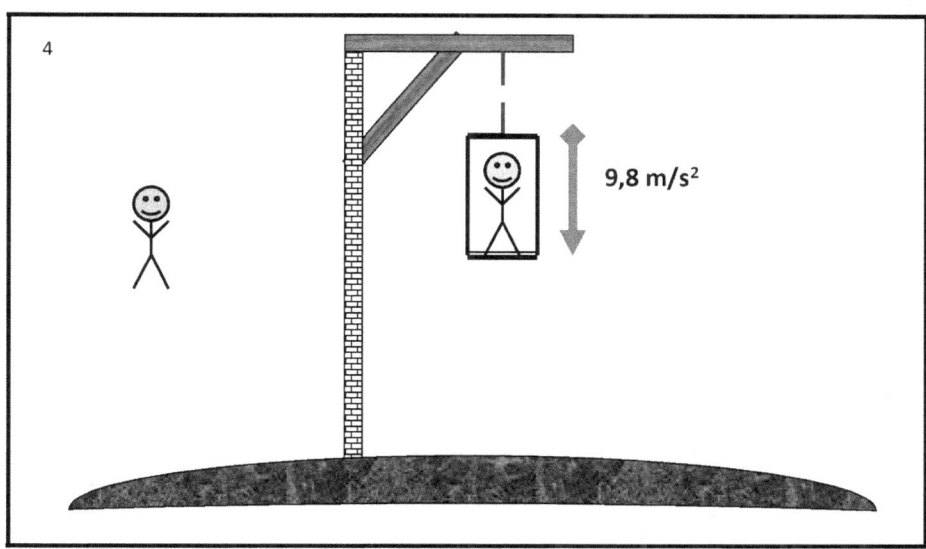

In figure 4, a portion of the earth's surface is shown, a strong vertical support on which a horizontal beam is fixed.

The elevator is roped to the girder. The rope is broken. For our consideration, it is not important whether the elevator was in motion or at rest at the time the rope broke. What is important is that the elevator will begin to fall toward the earth's surface, and it will move at an acceleration of nine whole eight tenths of a meter per second squared. The reason for this fall with acceleration is that the elevator, and the passenger in it, are in the gravitational field of the Earth, and experience the action of the force of the Earth's gravitational attraction. The elevator has no windows and the passenger in the elevator cannot know that it is moving with acceleration. The passenger in the elevator is in a state of weightlessness. The passenger in the elevator will be convinced that he is in a state of rest or uniform rectilinear motion, and no forces acting on him that cause acceleration. A second observer is located outside the elevator and sees that the elevator is moving with acceleration. The observer outside the elevator cannot convince the passenger inside the elevator that it is moving with acceleration towards the earth's surface.

It should be noted that similar thought experiments with elevators were conducted by Einstein to clarify the nature of inertial and non-inertial reference frames. These thought experiments helped Einstein define the Equivalence Principle.

Einstein used **the equivalence principle** to create the General Theory of Relativity.

General Relativity is a Theory of Time and Space. The General Theory of Relativity shows what the laws of mechanics are, and how the laws of mechanics work in non-inertial frames of reference. Non-inertial reference systems are those coordinate systems that are in a state of motion with acceleration. Modern physics and Einstein claim that accelerated motion is absolute, and thus differs from relative motion. The difference between absolute motion with acceleration on the one hand, and relative uniform motion on the other hand, is a very big problem that does not allow to create the General Theory of Relativity. The problem is solved by the Principle of Equivalence

The laws of relative uniform motion are a principle in the Special Theory of Relativity. From the history of physics, we know that Einstein first created the Special Theory of Relativity, then he created the General Theory of Relativity.

Special Relativity, like General Relativity, is a Theory of Time and Space. But unlike General Relativity, Special Relativity shows what the laws of mechanics are, and how the laws of mechanics work, in inertial frames of reference. Inertial reference systems are those coordinate systems that are in a state of rest or in a state of uniform rectilinear motion.

On July 11, 1923, Albert Einstein gave a speech in Gothenburg, before the meeting of natural scientists from the Nordic countries, on the topic: "Grundgedankenund und probleme der Relativatatstheorie".

The report was published in the book: "Les Prix Nobel en 1921-1922" Stockholm, Imprimerie Royale, PA Norstedt & Soner.

In this report, Einstein says:

"In classical mechanics, the distinction between accelerated and unaccelerated motions is absolute. There are only relative velocities depending on the choice of inertial frame, and accelerations and rotations are absolute, independent of the choice of inertial frame."

More than a hundred years ago, Einstein drew the attention of researchers to the essential difference between relative motion and absolute motion. The difference between absolute motion and relative motion is an obstacle to creating a General Theory of Relativity. Einstein tried to solve the problem by equating absolute motion with acceleration to relative motion with constant velocity. Philosophically speaking, this is a mistake. Einstein should have gone the other way, namely, to equate relative motion at constant velocity with absolute motion at constant acceleration. For this to happen, Einstein must represent, show, express relative motion at constant velocity by absolute motion at constant acceleration.

Einstein used the Equivalence Principle to equate absolute motion with acceleration, which is a principle in General Relativity, to relative motion, which is a principle in Special Relativity.

This is what Einstein says in the book "Evolution of Ideas in Physics":

"True relativistic physics must apply to all Coordinate Systems, and therefore also to the special case of an Inertial Coordinate System. The new **generalized** laws , valid for all Coordinate Systems , **must be** reduced to **the familiar old** laws **, in the special case** of an inertial system."

The blue text is:

"The new ones laws **valid** for all Coordinate Systems **are** reduced to laws , of an inertial system."

According to Einstein, **the new laws of physics** are true in coordinate systems that move with acceleration.

The principle of equivalence is used to bring absolute motion into relative motion, but this is not enough. Another very important fact is used.

An Inertial Coordinate System that enters a gravitational field begins to move with acceleration, but for the observers who are on that Inertial Coordinate System, nothing changes.

Observers do not feel the motion with acceleration. Observers are convinced that their coordinate system continues to be inertial and that it continues to move uniformly and in a straight line.

This is what Einstein says in the book "Evolution of Ideas in Physics":

"But for such a description, we have to account for gravity, building, so to speak, the bridge that makes it possible to go from one Coordinate System to another. The gravitational field exists for the external observer, but it does not exist for the internal observer."

And then:

"But the bridge, that is, the gravitational field, which makes possible the description in two different coordinate systems, rests on one very important pillar: the equality of heavy and inertial mass. Without this guiding thread, which has gone unnoticed in classical mechanics, our present rationale would be completely wrong".

The equality of heavy and inertial mass and the motion of an inertial frame of reference in a gravitational field are two of Einstein's wonderful ideas. Einstein used these two ideas to reduce absolute motion with acceleration to relative inertial motion. This is the path Einstein took and thus created the General Theory of Relativity.

From a philosophical point of view, Einstein's method suffers serious criticism. Einstein should have done the exact opposite, namely, tried to reduce relative inertial motion to absolute motion with acceleration.

In the hypothesis I present, you and I will do exactly that.

For this purpose, we will analyze basic physical laws and draw conclusions about the essence of these laws.

4. NEWTON'S FIRST LAW.

In 1868, Newton published the book

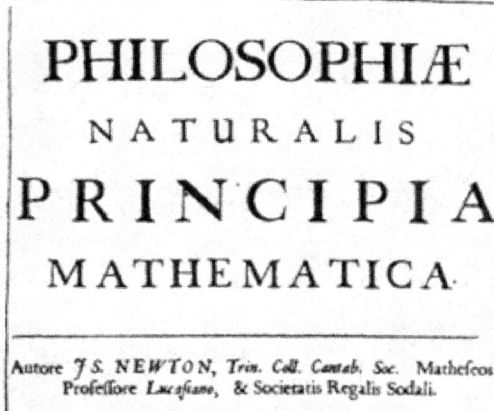

in which the basic laws of Physics are defined. The title of the book:

PHILOSOPHIAE NATURALIS PRINCIPIA MATHEMATICA

,

is translated into Slavic-Bulgarian Cyrillic, as follows:

„Математически принципи на физиката"

Newton's laws are studied in school and are known as "Newton's Three Laws".

In Latin, Newton's first law is written as follows:

„Corpus omne perseverare in statu suo quiescendi vel movendi uniformiter in directum, nisi quatenus illud a viribus impressis cogitur statum suum mutare"

The translation from Latin, into Slavic-Bulgarian Cyrillic, is written as follows:

„Всяко тяло продължава да запазва своето състояние на покой или равномерно праволинейно движение, докато и доколкото, то не е принудено да промени това състояние, от приложените сили"

The Latin to English translation is most likely spelled like this:

> "Every body continues to be held in its state of rest, or uniform and rectilinear motion, until and insofar as it is compelled by applied forces to change this state."

From Latin to Russian, there is a translation made by Academician Krylov in the book:

> ИСААК НЬЮТОН
>
> «МАТЕМАТИЧЕСКИЕ НАЧАЛА НАТУРАЛЬНОЙ ФИЛОСОФИИ»
>
> ПЕРЕВОД С ЛАТИНСКОГО И КОММЕНТАРИИ А.Н. КРЫЛОВА

The translation in Russian is written like this:

> "Всякое тело продолжает удерживаться в своем состоянии покоя или равномерного и прямолинейного движения, пока и поскольку оно не понуждается приложенными силами изменять это состояние"

Newton's first law:

"Any body continues to preserve its state of rest or uniform rectilinear motion, until and in so far as it is compelled to change that state by applied forces."

Quite deliberately I show the translation from Latin, in different scripts.

The reason is that what Newton says is very important. The way he says it is important.

Namely:

Newton's first law consists of two parts. The first part of Newton's law determines the state of the body in space and time when no **"force is applied" to the body** . Newton claimed that when on the body **it does not act "applied force"** , the possible state of the body is either rest or uniform rectilinear motion. Newton does not explain how rest or motion occurs. For Newton, the fact that these two states remain constant both in time and in space is important. The method of saving both states is the same. This means that the reason for maintaining the state of rest or the state of motion is the same. When **the cause of preservation** of these two different states is the same, then removing the cause of preservation will change the rest or the motion in the same way.

We must remember that the specific reason for the conservation of rest or motion, according to Newton, is **the absence** of an **"applied force."**

an **"applied force"** occurs , the state of rest or motion changes. In this way , Newton confirms the fact that **the reason for maintaining** the state of rest or motion is **the absence of the action of "applied force"** .

Newton's first law laid the foundations of the science of physics. From a philosophical point of view, Newton's first law has been heavily criticized. Criticism is related to the essence of the phenomenon of movement, and the essence of the phenomenon of rest:

Newton's first law does not distinguish between the state of rest of a body and the state of uniform rectilinear motion of the same body. To put it briefly and clearly, according to Newton's first law, the state of rest is identical to the state of motion, provided that the motion is uniform and in a straight line.

In science, philosophy, the phenomenon of motion and the phenomenon of rest are fundamentally different, and these phenomena have different essences. The identity of these fundamentally different phenomena creates problems for all modern physics. These problems can be specified in a variety of divisions of physics. A typical example in this regard is the Special Theory of Relativity. It is about the paradox of the twins. The twin paradox, which was defined by Einstein, states that when one of two twins moves uniformly and in a straight line relative to the other twin, the moving twin ages more slowly because time **slows down.** The only reason for the time delay is the fact that this twin is in a state of relative motion relative to the other twin. This hypothesis is funny, interesting, paradoxical, easy to remember, and arouses interest in a huge part of the readers. But I want to point out right away that the real paradox of twins is not the fact that there is a difference in the age of the twins. The true twin paradox boils down to the fact that each twin can claim to be aging more slowly and staying younger, while the other is aging faster. The reason for this misunderstanding is Newton's first law. I emphasize once again that Newton's first law does not distinguish between the state of rest and the state of uniform rectilinear motion.

See figure 5.

In figure 5, two twins and one platform are shown. The platform has wheels and can move. The twin who is on the right side of the figure has stepped on the platform. The platform, along with the twin on it, moves from left to right, uniformly in a straight line, at some speed. The direction and magnitude of the velocity is shown by a blue arrow. The twin on the platform says to the other:

"I move toward you, steady and straight, and I age more slowly."

But the other twin, who is located on the left side of the figure, objects:

"Oh no, what you say is not true, I am moving towards you. I am watching you carefully and I see that you are moving away from me at a constant speed".

The right twin responds:

"I am on a platform, and the wheels of that platform are turning, therefore I am in motion relative to you."

Thus, the dispute appeared to be already settled, in favor of one twin? Yes, it is solved, but the conditions of the experiment are violated. We are conducting an experiment which, by condition, aims to prove only and only relative, uniform, rectilinear, motion of the twins relative to each other. The wheels of the platform rotate, and their rotational motion is not uniform, it is not rectilinear. According to modern physics, the rotational motion of the wheels is absolute, and they must be excluded from the experiment we are conducting. The paradox of the twins refers, only and only, to a **state of relative motion, at a constant speed, in a straight line**.

The real experiment will look like this.

See Figure 6.

In figure 6, the two twins and two platforms are shown. The twins are on the platforms. Platforms don't have wheels because they are in outer space. The two platforms and the twins are in a state of weightlessness. The right platform, along with the twin on it, moves in a uniform straight line. The blue arrow shows the direction of the velocity and the magnitude of the velocity. It's deserted, completely empty, and the twins can determine speed relative to each other just by watching each other. Under these conditions, each of the twins can claim that he is moving while the other is at rest.

Each of the twins can use measuring devices to determine the relative speed of the other twin. For example, modern laser speed meters can be used.

See Figure 7.

Figure 7 shows the twins using laser speed meters. The red, thin arrows are laser light beams. In this case, each of the twins will be measured to be moving uniformly and in a straight line relative to the other twin. The velocity measured by the twins will be the same, but the direction of the velocity they measure will be opposite.

The right twin will claim to be moving from left to right, the left twin will claim to be moving from right to left.

The two blue arrows indicate the direction of the measured velocity. The length of the arrows indicates the magnitude of the measured velocity.

Pay special attention to the fact that the size of the arrows are the same, but the directions are diametrically opposite.

Placed in these conditions, the twins cannot determine which of the two is at rest and which is in motion. Here is another paradox. We see that the paradox of the twins consists of two parts, which are two fundamentally different paradoxes.

The first paradox is that one twin ages faster than the other twin. This is Einstein's paradox.

The second paradox is that it is in principle impossible to prove which of the two twins is at rest and which is in a state of uniform rectilinear motion.

From a philosophical point of view, the second paradox is extremely interesting, and is of particular importance. It's called **the Paradox of Motion and Rest.** The Twin Paradox, which was pointed out by Einstein, is a special case of the **Paradox of Motion and Rest.**

The only reason for the appearance and existence of **the Paradox of Motion and Rest** is that Newton's first law is defined in such a way that it does not distinguish between the state of rest and the state of uniform rectilinear motion. **The paradox of motion and rest** is like an evil demon living in the foundations of modern physics. This demon influences all human science.

5. NEWTON'S SECOND LAW.

In Latin, Newton's second law is written as follows:

> „Mutationem motus proportionalem esse vi motrici impressae et fieri secundum lineam rectam qua visilia imprimitur".

In Slavic Bulgarian Cyrillic:

> „Изменението на количеството на движение, е пропорционално на приложената движеща сила и се извършва по тази права по която тази сила действа"

In English:

> "The change in momentum is proportional to the applied driving force and occurs in the direction of the straight line along which this force acts"

In Russian:

> „Изменение количества движения пропорционально приложенной движущей силе и происходит по направлению той прямой, по которой эта сила действует"

Newton's second law:

"The change in the amount of movement is proportional to the applied driving force and is carried out according to the right on which this force acts" .

In his magnum opus, Philosophiae Naturalis Principia Mathematica, Newton defined the second law of physics in which he showed the relationship between physical quantities. The first quantity is **the amount of movement** , the second quantity is **the applied driving force** . The relationship between the **amount of movement** and the amount of **applied driving force** is reduced to two specific phenomena.

The first phenomenon is **proportionality** between amount of movement and applied force.

The second phenomenon is **a change in the amount of movement** .

Newton means that the amount of motion is directly proportional to the force and is directly proportional to the driving force.

The way it is stated, the second law of physics indicates that, for Newton, **the applied driving force** is the phenomenon that **causes the phenomenon of change** of **momentum** to occur . Note the fact that, said in this way, it indicates the presence of four different physical quantities.

The first is applied force.

The second is a driving force.

The third is amount of movement.

The fourth is a change in the amount of motion.

The new physical quantities are four, but for Newton the most important thing is that **the force causes the change** in the amount of motion to appear . This fact is confirmed in the second part of the definition of physical law, in Latin:

> "...et fieri secundum lineam rectam qua visilia imprimitur".

In Slavic Bulgarian Cyrillic :

> „...и се извършва по тази права по която тази сила действа".

In English:

> „...and occurs in the direction of the straight line along which this force acts"

In Russian:

> „...и происходит по направлению той прямой, по которой эта сила действует"

Translation from Slavic-Bulgarian Cyrillic to another language:
"...and it is done by that right by which that power acts" .

Newton, briefly and clearly, says that **the change** in the amount of motion takes place in a straight line and has a direction. The direction of the change in the amount of motion coincides with the direction of the acting force. That being said, it is extremely important.

Newton's definition is perfect. I say this because in modern physics, Newton's definition is presented in another way, and perfection disappears.

In modern physics, Newton's second law is written as:

"Force is equal to the product of the mass of the body times the

acceleration of the body."

Defined in this way, Newton's second law suffers serious criticism, from the point of view of science Philosophy. Philosophical criticism is in relation to the subordination of the three physical quantities which represent three different phenomena in the One Infinite Reality.

The three phenomena are: Force, Mass, Acceleration.

See Figure 8.

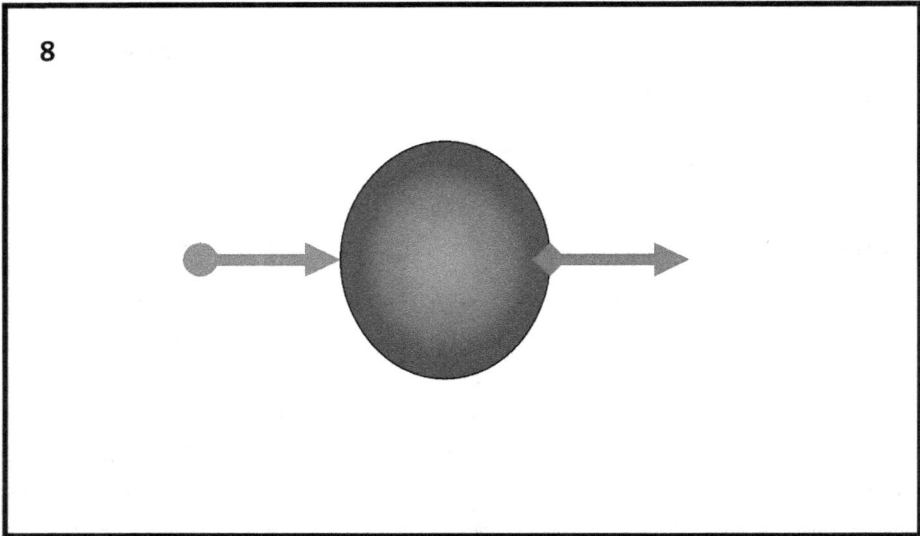

In figure 8, a sphere is shown which has a certain mass. The size of the mass in the specific case does not matter. A force is applied to the sphere. The force is shown with a red arrow. The length of the red arrow indicates the magnitude of the force. Under the action of the red force, the sphere moves with acceleration. Acceleration is shown with a green arrow. The length of the green arrow indicates the magnitude of the acceleration. The magnitude of the force acting on the sphere can be very different. If we use twice the force, then the acceleration of the sphere will be twice as great.

See figure 9.

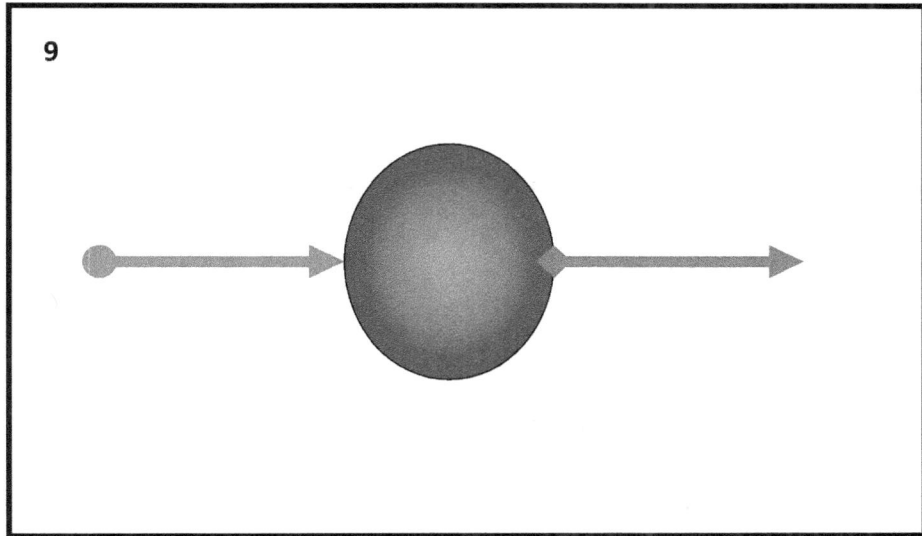

In figure 9, it is shown that the red force is twice as large compared to the force in figure four, then the acceleration is also twice as large. The green arrow shown in figure five is twice as large as the green arrow in the previous figure four.

We are able to change the size of the sphere as well. If we use twice the size of the sphere, and do not change the magnitude of the force, then the acceleration will be twice as small.

See Figure 10.

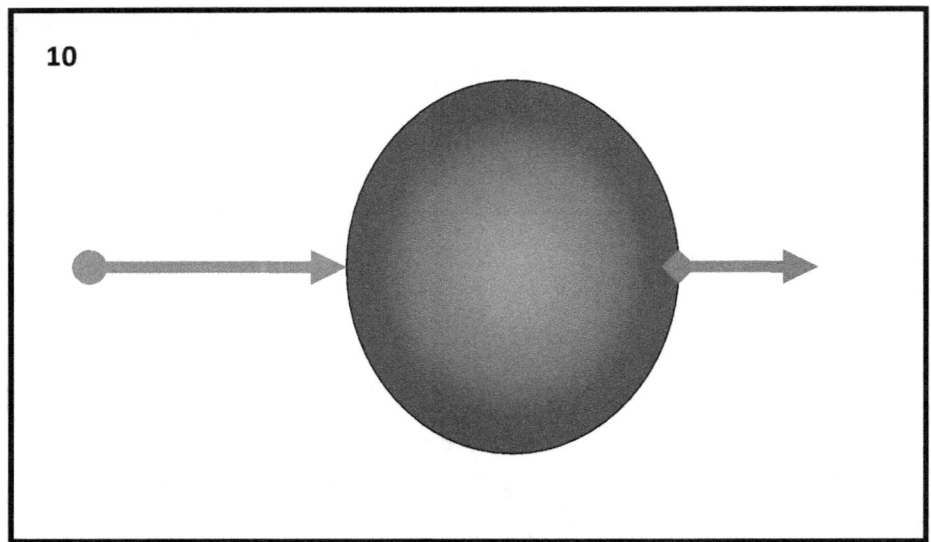

In Figure 10, a twice as large sphere is shown that is twice as heavy. The red force is not changed, but the acceleration, which is the green arrow, is twice as small, compared to the previous figure five.

We are able to make a variety of combinations between force, sphere weight and sphere acceleration. All possible combinations between these three physical quantities will be in agreement with Newton's second law as represented by modern physics, namely:

The force is equal to the product of the mass of the sphere times the acceleration of the sphere.

The philosophical question to Newton's second law is:

Which of these three physical quantities is primary?

Different answers are possible.

The first of the possible answers is that the Force is primary. Because if we observe a sphere on which no force is applied, the sphere will not be moving with acceleration, the sphere will be at rest. We apply a force to the sphere, and then an acceleration of the sphere occurs. Therefore, force is the thing that must appear first in order for acceleration to appear second. Force causes acceleration to occur.

But here philosophy immediately asks the next question, namely:

How does power appear?

The answer is that in order for a force to appear that can act on the sphere, some movement is necessary. The movement can be uniformly rectilinear or accelerating. It could be another sphere moving uniformly in a straight line, or moving with acceleration, towards the sphere we are experimenting with.

See figure 11.

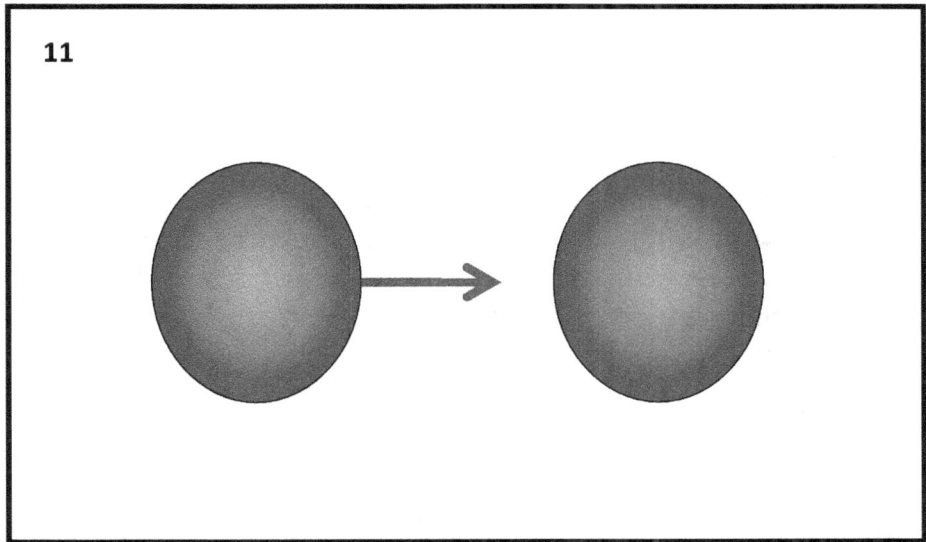

In Figure 11, two spheres are shown. The right one is at rest. The left sphere is moving towards the right with some speed. The direction of the velocity and the magnitude of the velocity is shown with a blue arrow.

See figure 12.

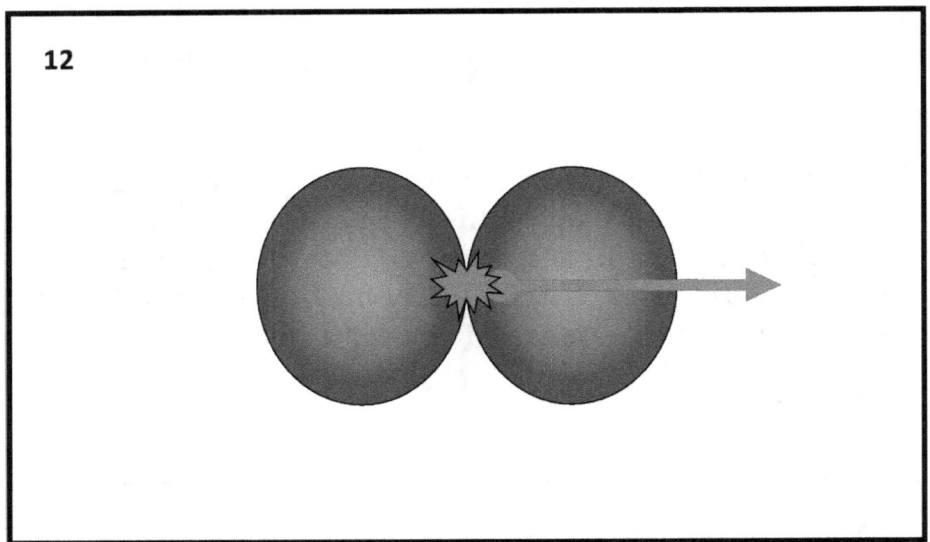

In figure 12, the impact between the two spheres is shown. At the moment of impact, accelerations occur between the atoms that make up the spheres. The red burst shows the accelerations that occur at the quantum level. These accelerations give rise to the force that begins to push the sphere with which we are doing experiments.

But then, maybe acceleration is primary?

But we must not forget that in order for any acceleration to occur, some action of force is always necessary, which is applied to some body possessing some mass. Then, we may conclude that acceleration is not primordial.

A third possible answer is that the mass of the sphere is a primary physical quantity. Because if we change the mass of the sphere but keep the magnitude of the acting force, the acceleration will change. We can conclude that the change in the mass of the sphere is the cause of the change in acceleration.

But to co-create the accelerating motion of the sphere, the action of a force is necessary. If there is no force acting, the sphere will not move with acceleration.

A closed circle is obtained. Each of these physical quantities is the cause of the appearance of the other two, and this happens through a rigorously proven physical dependence. This physical dependence is called Newton's second law.

Modern physics is unable to determine which of these three physical quantities is primary. When the primacy of one of the three quantities is proven, then it will be the reason for the appearance of the other two physical quantities. For now, this has not been done.

This is a serious problem of modern physics that affects all human science.

The reason for this problem is that the modern definition of Newton's second law differs from the original definition that Newton proposed. At the beginning of this chapter I showed that according to Newton:

The "**applied driving force**" causes a "**change**" in the "**amount of motion**" to occur.

This is very important and must be remembered.

6. NEWTON'S THIRD LAW.

Newton's third law written in Latin:

> „Actioni contrariam semper et aequalem esse reactionem: sive corporum duorum actiones in se mutuo semper esse aequales et in partes contrarias dirigi"

Written in Slavic Bulgarian, Cyrillic:

> „Действието винаги е равно и противоположно на противодействието, иначе казано взаимодействията на две тела, едно върху друго, по между си, са равни и са насочени в противоположни посоки"

Written in Russian:

> „Действию всегда есть равное и противоположное противодействие, иначе — взаимодействия двух тел друг на друга между собою равны и направлены в противоположные стороны".

Written in English:

> „An action always has an equal and opposite reaction, otherwise the interactions of two bodies against each other are equal and directed in opposite directions".

Translated from Slavic Bulgarian Cyrillic, into another language:

"The action is always equal and opposite to the counteraction, in other words the interactions of two bodies, one on the other, between themselves, are equal and directed in opposite directions"

The law is defined concisely and clearly.

From a philosophical point of view, Newton's third law has suffered serious criticism.

There are no restrictive conditions in the definition of the law. Limiting conditions indicate when the law applies and when the law does not. The absence of restrictive conditions gives reason to some researchers to claim that Newton's third law ranks as a physical principle.

The absence of a definitional area that shows how the law works is a prerequisite for the existence of speculations that make it difficult to properly understand the nature of the law. In this way, the view appears that the force of counteraction does not exist, and that the force of counteraction is a fictitious force.

The essence of the law is revealed through figures.

See Figure 13.

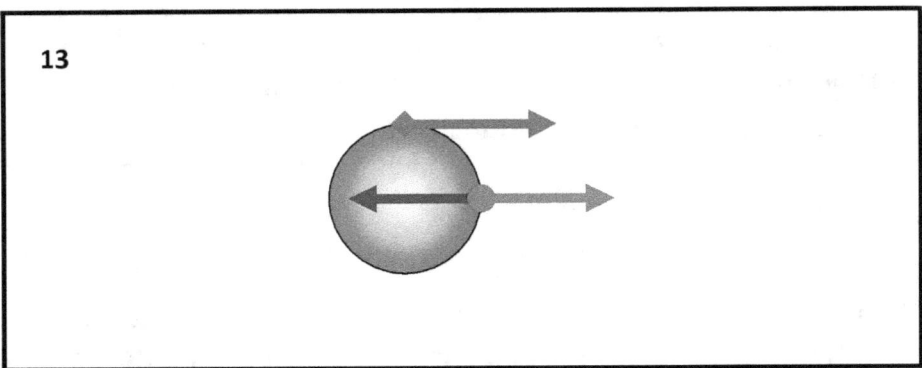

In figure 13, a sphere is shown, and the forces acting on the sphere. A red force is applied to the sphere, which pulls the sphere to the right, and a blue force, which opposes the red one. The red force pulls on the sphere and the sphere begins to move with acceleration. Acceleration is shown with a green arrow. The direction of the acceleration coincides with the direction of the pulling red force.

An acting force can be a pushing force. It depends on the point of application of the force.

See figure 14.

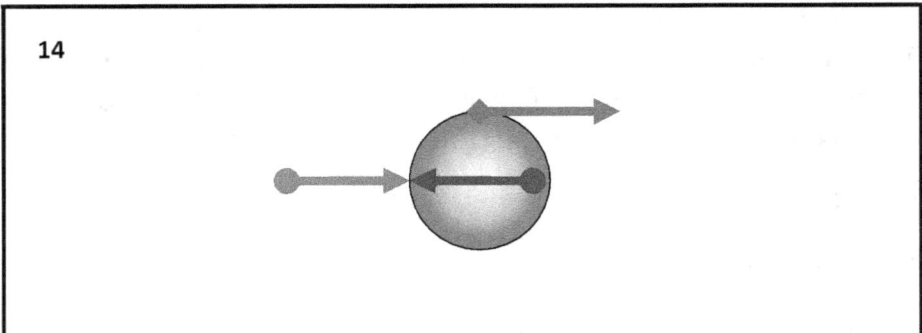

Figure 14 shows a red pushing force and a blue force that opposes the red one. The green arrow shows the direction of the acceleration. A case of central force action is also possible.

See Figure 15.

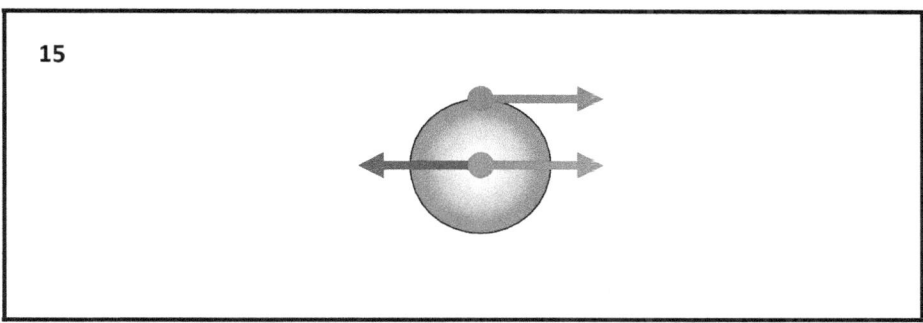

In figure 15, a centrally acting red pulling force is shown, and a blue force counteracting the red one. The green arrow shows the magnitude and direction of the acceleration.

Some reader may ask: Why am I describing these elementary things in such detail?

My answer is this:

Because this book is for people who do not have a special education in Physics.

Because these things are very important and must be understood properly.

Because I've taught physics, both to children and adults, and they all claim to know Newton's third law, and they're convinced they understand it. And as the conversation goes on, some of them conclude that the counterforce doesn't exist, that the counterforce is a fictitious force, put there for convenience.

Some of my students, after looking at figure 15, say the following:

"Blue power is equal to red power, and blue power is the opposite of red power. Then these two forces cancel each other out. Therefore, the sphere cannot move with acceleration. If the sphere is moving with acceleration, then the blue force is fictitious. Blue does not exist. The countermeasure does not exist. Only the red pulling force continues to act, and then, from Newton's second law, it follows that the sphere moves with acceleration."

The question arises: What grounds such a conclusion?

The answer lies in the fact that in the science of physics there are two large, distinct divisions. These are called dynamics and statics. When conducting physical thought experiments, one must always consider which of these two branches of physics the particular experiment is about.

See Figure 16

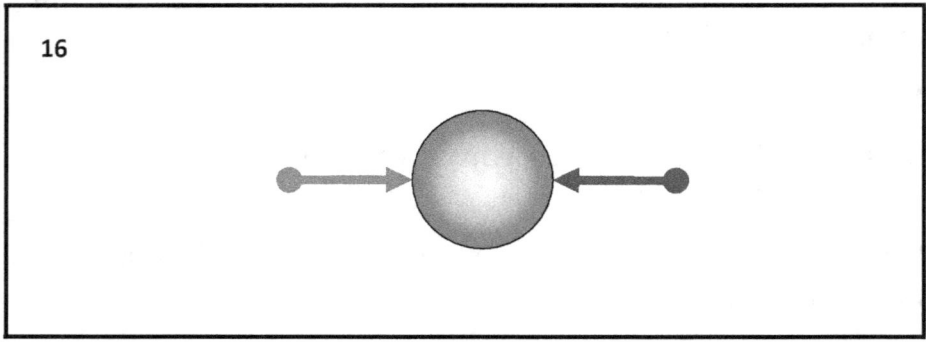

Figure 16 shows a sphere and two forces simultaneously acting on the sphere. The blue force is equal to the red force, and both forces are directed against each other. The blue and red forces cancel each other out, and the sphere is either at rest or in uniform rectilinear motion. This is a classic experiment from the statics section of Physics. The figure twelve shown is very similar to the figures thirteen, fourteen, and fifteen. The essential difference between the two figures is that the application points of the forces are two different ones. The blue power has its own application point, which is different from the application point of the red power. When we analyze Newton's third law, the action force and the reaction force have the same application point, which is shown in figure eleven. This fact is very important, and to understand it, we have to read what Newton says in his book "Mathematical Principles of Physics".

"If something presses on something else or pulls on it, then it itself is crushed or pulled by the latter. If one presses a stone with his finger, then his finger is also pressed by the stone. If the horse drags a stone tied to a rope, then, conversely (so to speak), it pulls on the stone with equal effort, because a taut rope, due to its elasticity, produces the same force on the horse to the stone, and on the stone to the horse, and as much as this rope prevents the horse from going forward, so much does it make the stone go forward' .

In Slavic-Bulgarian Cyrillic:

„Ако нещо притисне нещо друго или го дърпа, то самото то се смачква или издърпва от това последното. Ако някой натисне камък с пръста си, тогава неговият пръст също е притиснат от камъка. Ако конят влачи камък, вързан за въже, тогава, обратно (така да се каже), той се дърпа към камъка с еднакво усилие, защото опънато въже, поради своята еластичност, произвежда същата сила върху коня към камъка и на камъка към коня и колкото това въже пречи на коня да върви напред, толкова и кара камъка да върви напред".

In English:

„If something presses on something else or pulls it, then it itself is crushed or pulled by this latter. If someone presses a stone with his finger, then his finger is also pressed by the stone. If a horse drags a stone tied to a rope, then, back (so to speak), it is pulled towards the stone with equal effort, because the stretched rope, by its elasticity, produces the same force on the horse towards the stone and on the stone towards the horse, and as much as this rope prevents the horse from moving forward, so much does it impel the stone to move forward"

In Russian:

„Если что-либо давит на что-нибудь другое или тянет его, то оно само этим последним давится или тянется. Если кто нажимает пальцем на камень, то и палец его также нажимается камнем. Если лошадь тащит камень, при-вязанный к канату, то и, обратно (если можно так выразиться), она с равным усилием оттягивается к камню, ибо натянутый канат своею упругостью производит одинаковое усилие на лошадь в сторону камня и на камень в сторону лошади, и насколько этот канат препятствует движению лошади вперед, настолько же он побуждает движение вперед камня"

With the help of a few figures, I will show what is action and what is counteraction.

See Figure 17.

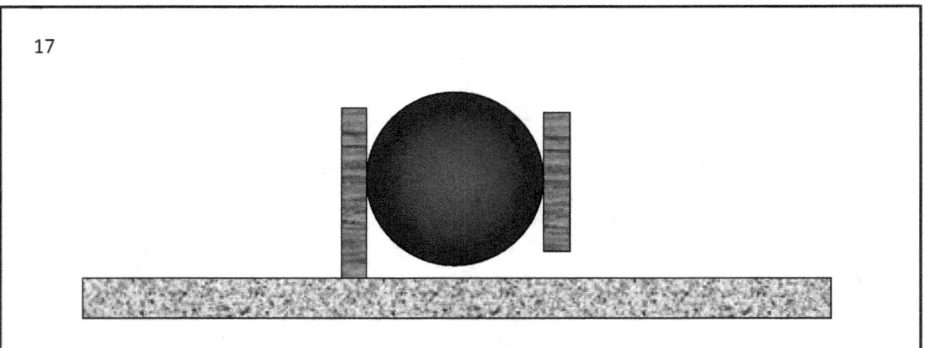

Figure 17 shows a blue rubber ball. The ball is located between two light boards, boards. The left board is firmly fixed on a heavy slab made of stone, granite. The right board is free and can be moved. We apply a force action on the right board.

See figure 18.

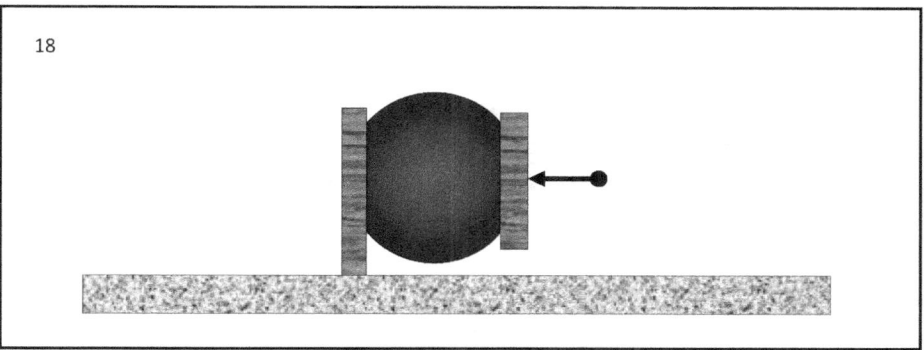

In Figure 18, it can be seen that the black force is applied to the right board. The board is placed to prevent the ball from popping. The force acts from right to left. The board presses on the rubber ball, and the ball deforms from right to left. Exactly the same deformation will occur on the left side of the ball. A board is placed there, which is firmly connected to the granite slab, and is immovable. Look at the figure. The ball is deformed on both sides equally. The right deformation is caused by **the action** of the right board, on the ball. The left warp is caused by **the counteraction** of the left board on the ball. I can say that this is a perfect classic experiment showing **action** and **counteraction** , in the statics section of the science of Physics. Let's check what Newton says in his great work "Mathematical Principles of Physics".

"If one presses a stone with his finger, then his finger is also pressed by the stone."

An experiment can be made that shows the action and counteraction in the dynamics section of the science of Physics.

See Figure 19.

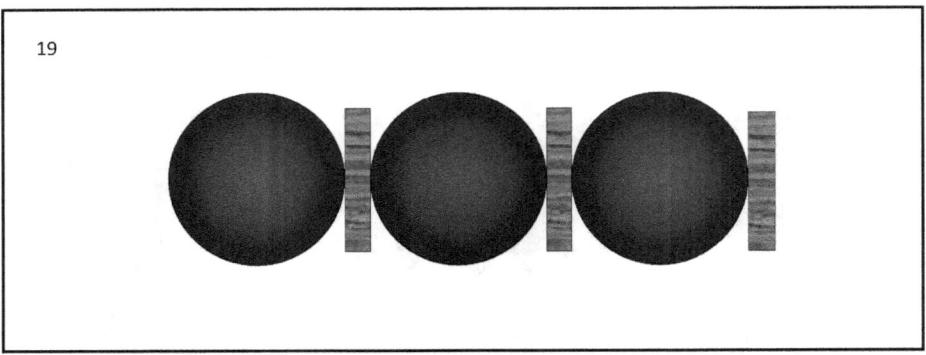

Figure 19 shows three blue rubber balls and three light boards made of wood. We apply force action.

See Figure 20.

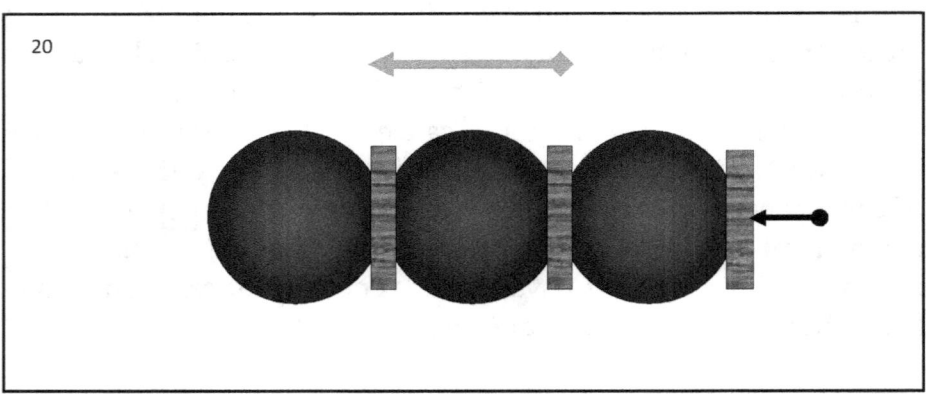

Figure 20 shows the balls, boards, and black force acting from right to left. The action of the black force forces the balls and boards to move with acceleration, from right to left. The green arrow at the top is the acceleration. Look carefully at the figure and you will understand **the action** and **counteraction** in the dynamics section of the science of Physics.

The left board and middle board can be removed. Not the rightmost one, because the ball will burst. By removing the two boards, the deformation of the three balls will not change. You already know why.

The essence of Newton's third law boils down to the following statement:

For every action of a force, there is an equal in magnitude and opposite in direction acting force.

The question arises:

What is the magnitude of these two forces, and how can we be sure that they exist and always act simultaneously?

We will do a thought experiment, and show and measure a real force acting on a sphere.

See Figure 21.

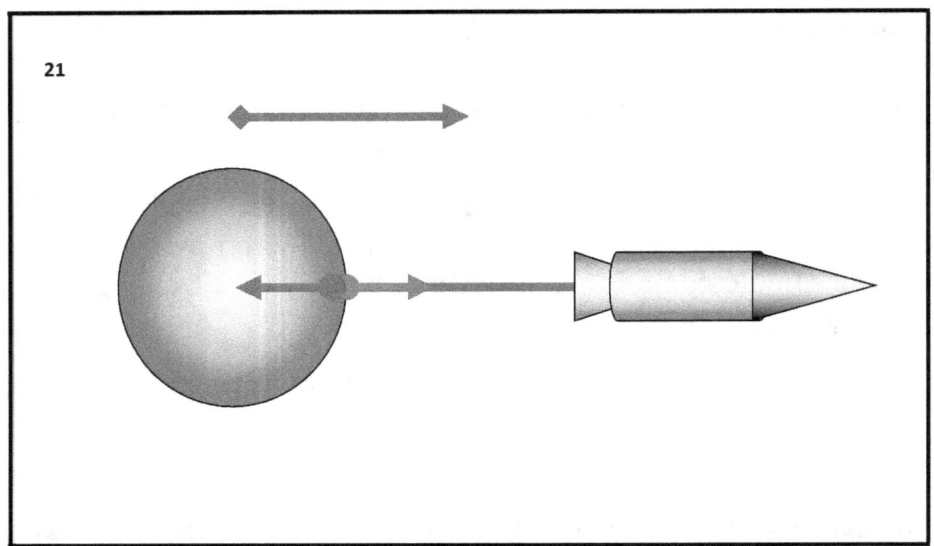

In figure 21, the sphere is shown, and a rocket is tied to the sphere with a rope. We start the rocket engine, the rocket pulls the rope, and the rocket starts pulling the sphere. The rocket acts on the sphere with some force. The sphere begins to move with acceleration. Acceleration is shown with a green arrow. The red arrow is the action force, the blue is the reaction force. The force of action and the force of counteraction must be measured. Forces are measured using a force meter.

See Figure 22.

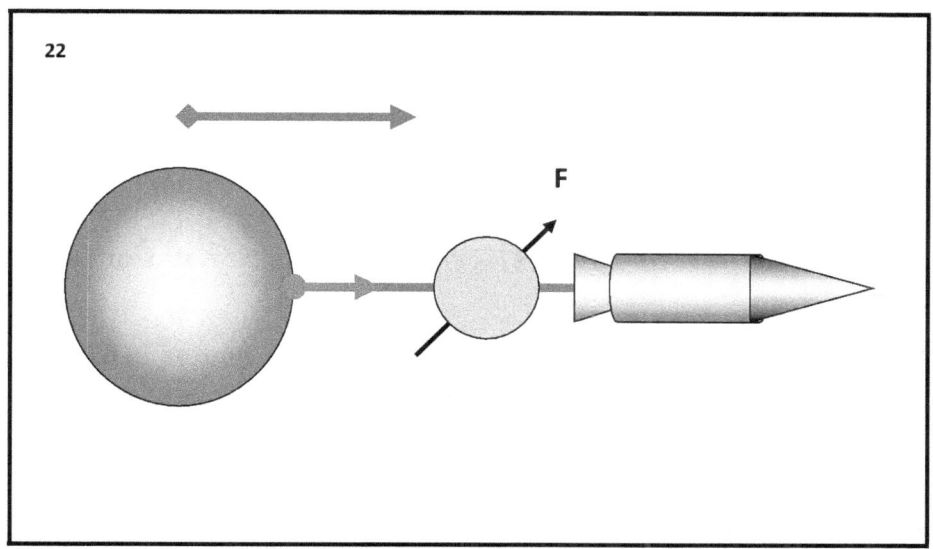

In figure 22, the sphere, the rocket and the rope between them are shown. A force meter is placed in the middle of the rope, which measures the action and counteraction. The red force is the force of action, the blue force is the force of counteraction. The green arrow shows the acceleration.

Figure twenty-two shows the essence of Newton's third law.

The experiment shown in figure eighteen proves and explains the existence of action and counteraction. Whenever we analyze Newton's third law, we must imagine the experiment shown in this figure, and the experiment with the three blue balls.

7. NEWTON'S LAW OF GRAVITATION.

According to modern physics, Newton's law of gravitation states that:

The force of gravitational attraction between bodies is directly proportional to the product of the two bodies, and inversely proportional to the square of the distance between the two bodies.

Put another way, the magnitude of the gravitational force with which two bodies are attracted to each other is equal to the mass of one body times the mass of the other body divided by the distance between the two bodies squared.

Newton's law of gravitation is written as:

$$F = \frac{M.m}{r^2}.G$$

Where:

F is the force of gravitational attraction between the two bodies.

M is the mass of the larger body.

m is the mass of the smaller body.

r is the distance between the centers of the two bodies.

G is the gravitational constant.

From a philosophical point of view, Newton's third law has suffered serious criticism.

Philosophical criticism is directed against the way in which the phenomenon of force is defined in modern physics. In modern physics, there are two different mathematical expressions for force. The two mathematical expressions were stated by Newton.

The first mathematical expression is represented by Newton's second law, which states that:

Force is equal to the product of mass and acceleration.

$$F = m.a$$

The second mathematical expression, represented by Newton's law, is the force of gravitational attraction.

$$F = \frac{M.m}{r^2}.G$$

The fact that there is equality between heavy and inertial mass, and Einstein's **principle of equivalence**, allows us to establish equality between these two mathematical expressions. It is obtained:

$$F = \frac{M.m}{r^2}.G = m.a$$

The possibility of writing this equality in this way, from a philosophical point of view, is a shortcoming of modern physics. Einstein's principle of equivalence legitimizes the mathematical expression for the equality of the two forces.

Einstein's equivalence principle plays an extremely important role in modern physics.

Einstein's principle of equivalence lies in the foundation of the General Theory of Relativity.

Einstein's Principle of Equivalence is a fundamental law by which human conceptions of the One Infinite Reality are created.

The equivalence principle is a paradigm in modern human science.

8. RELATIVE MOTION AT CONSTANT VELOCITY.

Einstein says that the constant speed of a test body depends on the choice of **inertial frame of reference.**

See Figure 23.

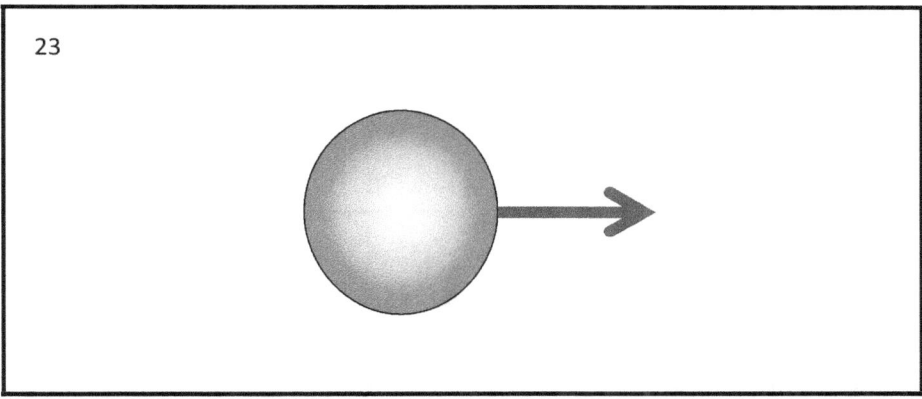

In figure 23, a sphere is given which **moves with a constant speed**. The blue arrow shows the direction and magnitude of the constant velocity.

From a physical point of view, the expression **moves at a constant speed** is incomplete and inaccurate because no numerical value of the velocity magnitude is given, and no coordinate system is given.

The phenomenon of a numerical value of **a magnitude** of constant velocity has a physical meaning only when the coordinate system relative to which the sphere moves at a constant velocity is indicated.

See figure 24.

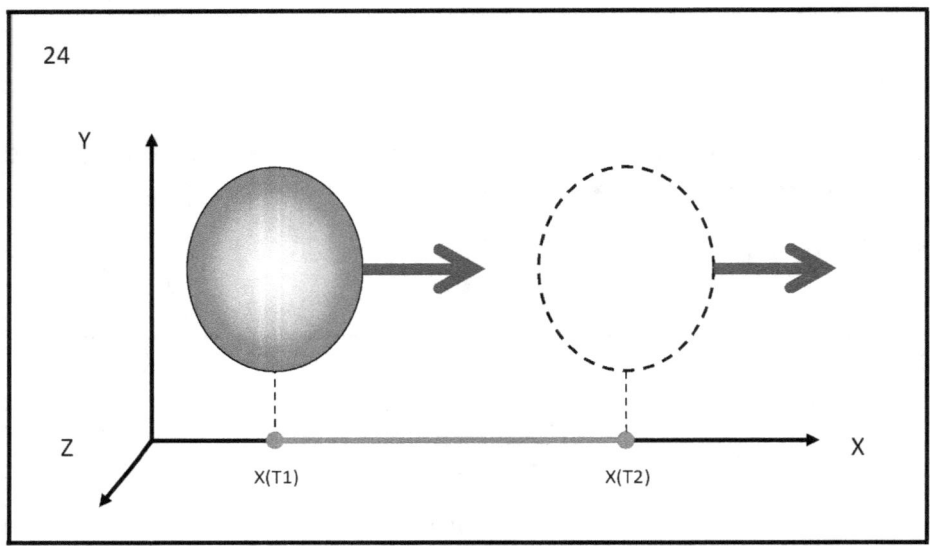

Figure 24 shows a coordinate system and a sphere moving at a constant speed relative to the coordinate system. Constant speed is shown with a blue arrow. In this coordinate system, the sphere moves some distance, in some time. The move is shown in red. When we divide the displacement by the time interval, we get the velocity of the sphere relative to this coordinate system. The length of the blue arrow indicates the magnitude of the constant velocity. The magnitude of the constant speed of the sphere depends on the state of motion or rest of any one specifically chosen inertial frame of reference. If we choose another inertial coordinate system, the velocity will be different.

See figure 25.

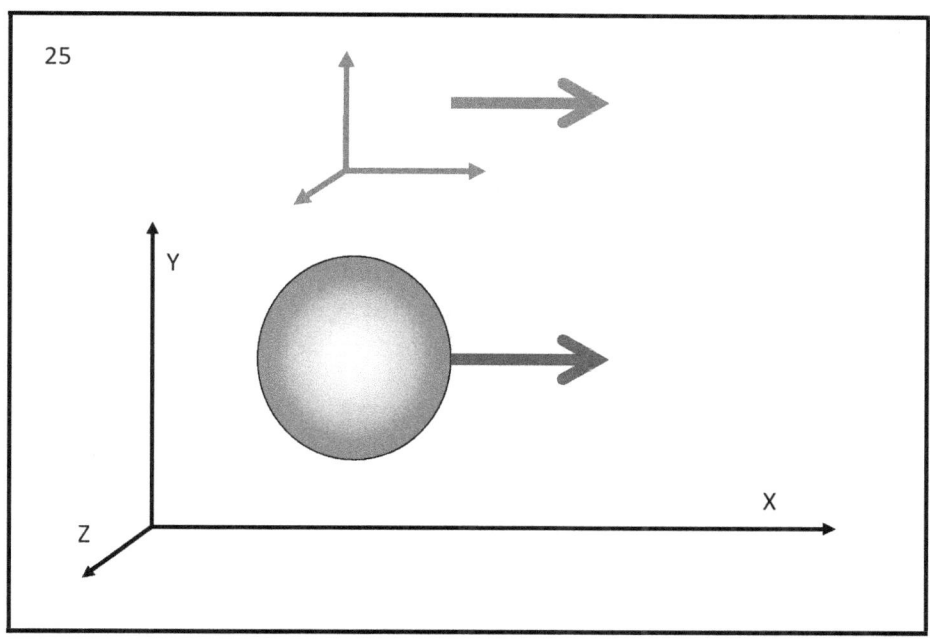

Figure 25 shows a large coordinate system made of black arrows, a sphere moving at a constant speed relative to the black coordinate system, and a small coordinate system made of green arrows. The green coordinate system moves at constant speed. The magnitude of the velocity and the direction of the velocity is shown by a green arrow. The green arrow is equal to the blue arrow. The sphere and the green coordinate system are moving, side by side, at the same constant speed, in the same direction. The sphere is then at rest relative to the green coordinate system.

The sphere is simultaneously in two states, namely, at rest relative to the green coordinate system, and in a state of motion, with constant velocity, relative to the black coordinate system.

The velocity of the sphere in the green coordinate system is zero, the velocity of the sphere in the black coordinate system is greater than zero.

When Einstein says that the constant speed of a test body depends on the choice of **inertial frame of reference,** he means what we have shown with the figures.

Relative constant velocity means dependent constant velocity .

The velocity dependence is relative to **the choice** of the coordinate system, and depends on the magnitude of the velocity with which **the selected** coordinate system is moving. **The choice** of a coordinate system relative to which the velocity **measurement is made is the choice** of another, different velocity.

Selection and measurement are forms of reflection realized by the subject who performs the particular experiment .

Find and see on the net: "Theory of reflection" by academician Todor Pavlov.

Each experimenter is a subject in relation to the object present in the experiment. When the subject first makes a choice about the state of the object, then the subject proposes a particular new state. In the experiment we are analyzing there are two specific states, namely rest or movement. The new state proposal is a convention proposal. A convention is a contract that states what is true and what is not true. The contract can be accepted by the other researchers, subjects. But it can also be rejected. This is called conventionality in science. Philosophically, conventionality is a huge problem in modern human science.

9. ABSOLUTE MOTION WITH CONSTANT ACCELERATION.

Albert Einstein says:

"accelerations and rotations are absolute, they do not depend on the choice of the inertial system".

What Einstein says is very important. It needs to be understood very well.

See figure 26.

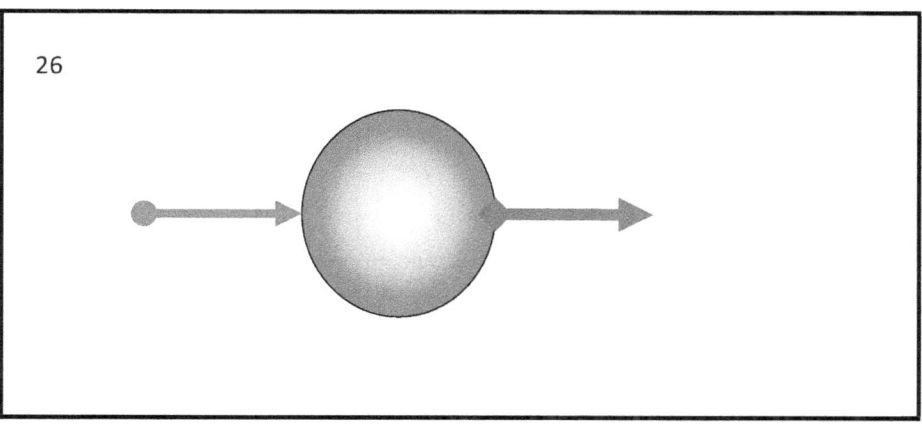

In figure 26, a sphere and two arrows are shown. The red arrow is a force pushing the sphere from left to right. Under the action of the red force, the sphere moves with acceleration, from left to right. The green arrow shows the direction and magnitude of the acceleration. No coordinate system shown. It is not necessary. Because the acceleration of the sphere is absolute, which means that the measurement of the magnitude of the acceleration can be done without the need for a coordinate system. This means that the acceleration of the sphere does not depend on the choice of coordinate system. We can choose any inertial coordinate system and measure the acceleration of the sphere relative to it. The magnitude of the measured acceleration will be the same, a constant.

See figure 27.

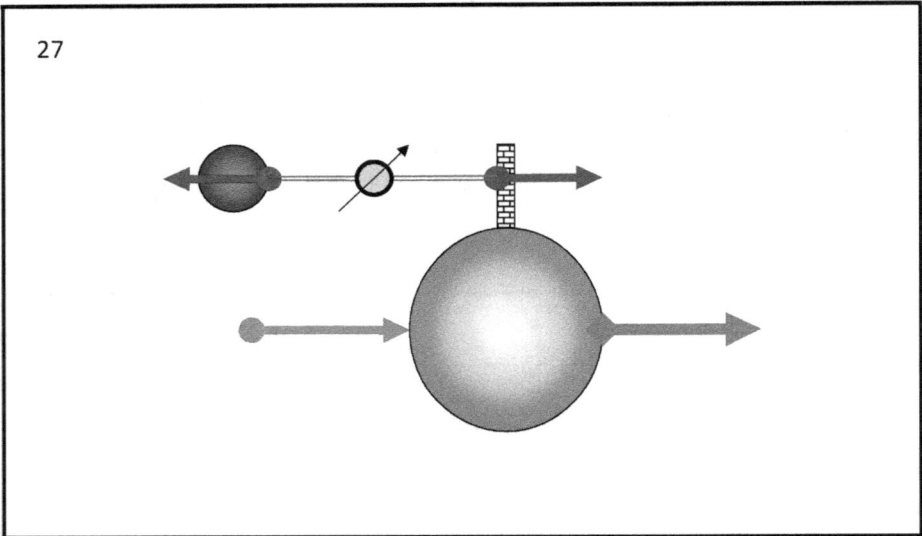

Figure 27 shows a red force pushing the sphere from left to right. Under the influence of the force, the sphere moves from left to right with acceleration. The direction and magnitude of the acceleration are shown with a green arrow. A retaining wall is made at the upper end of the sphere. A small red sphere is given which is tied to the wall with a brown rope. In the middle of the rope, a force measuring device, a force meter, is placed. The red sphere is a sample body that is selected with a reference mass. The wall pulls the small red sphere, with some force, which is shown by a purple arrow. In accordance with Newton's third law, the small red sphere counteracts the purple force, with a force equal in magnitude but opposite in direction. The countermeasure is shown with a blue arrow. The force meter measures action and counteraction.

The mass of the red reference sphere is known, the magnitude of the purple force acting on it has already been measured. Using Newton's second law, the acceleration of the small sphere is calculated. The calculated acceleration of the small red sphere is equal to the acceleration of the large sphere. This is only one way to determine the acceleration of the large sphere. This method is universal. It is possible to use different test bodies to be placed at different places on the large sphere. Through these test bodies, we can always measure the force of action and the force of counteraction, and thus determine the magnitude of the force acting on the specific test body, after which we calculate the acceleration.

No coordinate system is used to determine the acceleration. The method we used shows that the acceleration **does not depend** on the coordinate system, which is moving at a constant speed, or is in a state of rest.

This is why, Albert Einstein said:

"accelerations and rotations are absolute, independent of the choice of inertial frame."

See figure 28.

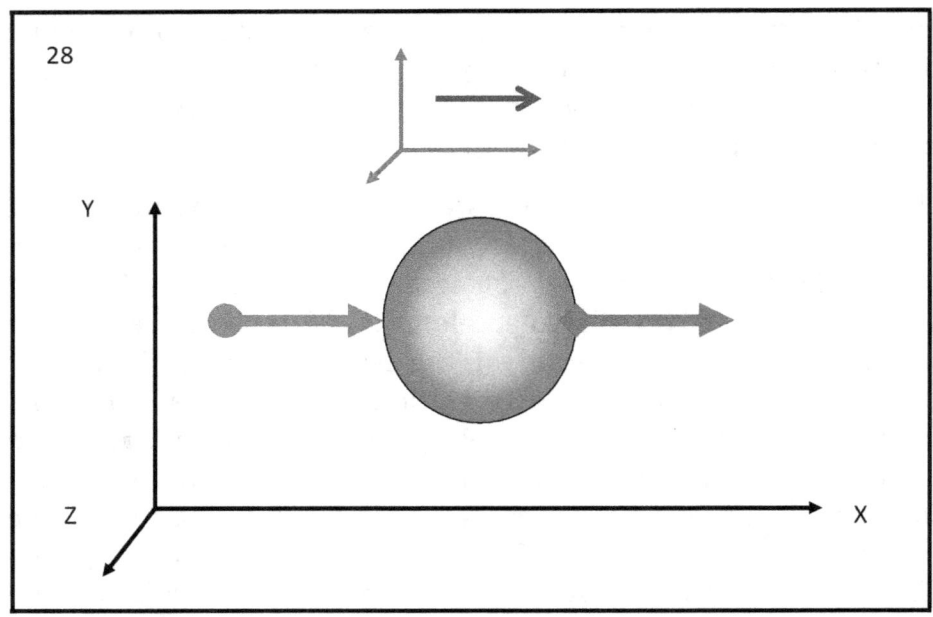

In figure 28, a coordinate system made of black arrows is given, which is at rest.

A small coordinate system is given, which is made with green arrows. The small green coordinate system moves relative to the large black coordinate system, at a constant speed, uniformly in a straight line. The magnitude of the velocity and the direction of the velocity in the green coordinate system are shown by the blue arrow.

Given a sphere on which the action of a red thrust is applied. Under the action of the red thrust, the sphere moves with acceleration. Acceleration is shown with a green arrow. The direction of the red force matches the direction of the green acceleration. The length of the green arrow indicates the magnitude of the acceleration.

The sphere moves with **the same acceleration** relative to the large black coordinate system and relative to the small green coordinate system. The big black one is at rest, the little green one is moving, but nevertheless, the acceleration of the sphere is the same for both coordinate systems. The reason for this equality is that the acceleration is absolute.

I have shown a detailed proof of this statement in The Paradox of the Rod. Part Six." Publishing house E.D.B. Amazon. This is a comic for children and adults, in which I have presented the basic laws of physics through drawings.

10. ATTRIBUTION OF TYPES OF MOVEMENTS.

*****

Philosophical explanations

The modern science of physics defines two basic types of motion, which are absolute motion and relative motion.

The concept of **absolute** and the concept of **relative** are philosophical categories. In human science, the relationship between these two categories is unclear. In the general case, the absolute and the relative are opposed, and placed in a position of antagonistic contradiction. This approach is wrong. The absolute and the relative are in a dialectical unity. The **absolute** category and the **relative category** are a pair of categories.

I propose to use the idea that the dialectical relationship between the category **relative** and the category **absolute** is as follows :

The absolute refers.

The relative becomes absolute.

In this way, they are included in the pairs of categories of Hegel's dialectic.

*****

Absolute motions are well known to modern physics. I have already said that according to Einstein, motion with acceleration and rotational motion are absolute motions. The relations between the different types of absolute movements are diverse, and it is necessary to be subjected to a general philosophical, dialectical analysis.

For this purpose, we will perform appropriate thought experiments.

See Figure 29.

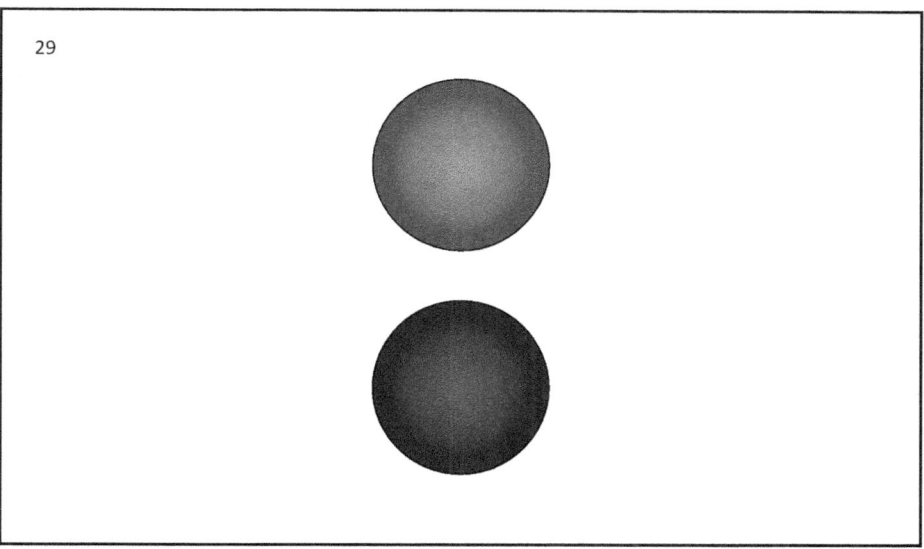

In Figure 29, two spheres are shown. Green sphere and blue sphere. The spheres are the same size and have the same mass. The two spheres are at **rest relative to each other**. No coordinate system is shown in the figure.

*****

Philosophical explanations:

When we, the subjects conducting the experiment, say " **at rest**

relative to each other ", it means that we, **the subjects**, do not need a coordinate system to prove the state of rest between the two spheres.

This means that **the objects** of the experiment, which are the two spheres, do not need a coordinate system to prove, show, establish, the state of rest of the two spheres.

No coordinate system is shown in the figure.

This means that the state of rest between the two spheres depends only and solely on the two spheres, and on **the relation** of one sphere to the other sphere. The physical conditions under which the relation between the two spheres takes place are pre-defined by the subject performing the experiment.

The concept of **attitude** is a philosophical category. The act of **relating** between the two spheres proves, shows, establishes the state of rest that objectively **exists** between the two spheres. The objective existence of the state of rest, under specific conditions, absolutizes the state of rest between the two spheres. The correct sentence is:

The two spheres are in a state of absolute rest **relative to each other.**

The state of absolute peace between the two spheres is possible through the relation, only and only, of one sphere to the other sphere, and vice versa.

*****

We, the subjects who perform the experiment, apply an action of force to the two spheres that are the subject of the experiment.

See Figure 30.

EINSTEIN'S THIRD MISTAKE

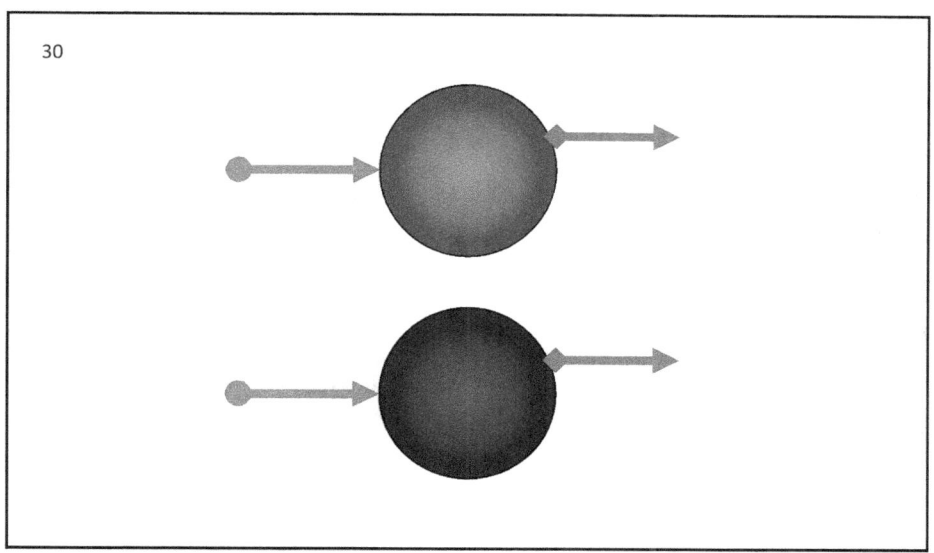

In Figure 30, it can be seen that two equal, red, pushing forces are applied to the two spheres. There is no coordinate system in the figure. The length of the two red arrows is the same.

The two pushing forces are applied simultaneously to both spheres. The two spheres simultaneously begin to move with acceleration. Acceleration is shown with green arrows. The acceleration of the two spheres is the same. The length of the green arrows is the same.

*****

Philosophical explanations:

From a philosophical point of view, both spheres are subject to experimentation. The researchers conducting the experiment are the subjects of the experiment. We subjects observe and analyze the movement of the spheres. Observing, measuring and analyzing are forms of **reflection** . **Reflection** is a philosophical category that we specified in the definitional framework. The subject's reflection of the object is always subjective.

See on the Internet: Academician Todor Pavlov, "Theory of reflection".

We have said that the two spheres are at relative rest relative to each other.

In the figure, two different phenomena are **observed and reflected at the same time.**

The first phenomenon is that the two spheres **move absolutely**, with the same **acceleration**, side by side, in the same direction.

The second phenomenon is that the two spheres are in a state of **relative rest** relative to each other. These are two different phenomena that are observed simultaneously.

We have already explained that to establish these two phenomena we do not need a coordinate system.

*****

I have already said that on July 11, 1923, Einstein gave a speech in Gothenburg, before the meeting of naturalists from the northern countries.

In this report, Einstein says:

"In classical mechanics, the distinction between accelerated and unaccelerated motions is absolute. There are only relative velocities depending on the choice of inertial frame, and accelerations and rotations are absolute, independent of the choice of inertial frame."

From a philosophical point of view, this statement of Einstein is subject to serious criticism.

EINSTEIN'S THIRD MISTAKE

The criticism boils down to the fact that in the experiment we are conducting, we observe the phenomenon **of relative rest** of two spheres that move with **absolute acceleration**.

A question arises:

Why, until now, in human science has not been specifically noted that there is a state of relative rest, between two things moving with absolute acceleration? This, in my opinion, is a fundamentally important phenomenon.

We will use this fact to create a hypothesis.

See Figure 31.

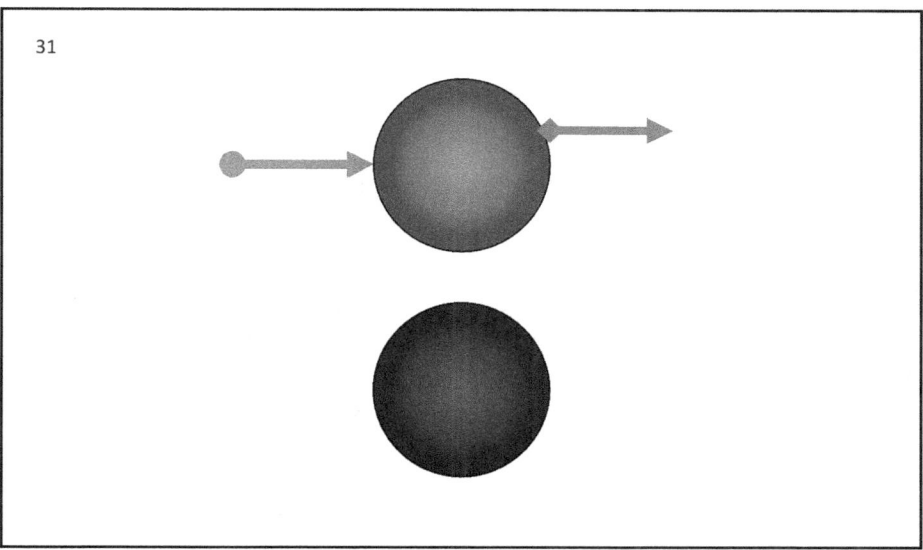

In figure 31, the two spheres are given. The blue sphere is at rest. A red thrust is applied to the green sphere. The red sphere begins to move with acceleration relative to the blue sphere. The direction of the acceleration is shown by a green arrow. The magnitude of the red force is such that the green sphere moves with an acceleration of one meter per second squared. The accelerating motion of the green sphere is done relative to the blue sphere. Proving the accelerating motion of the green sphere does not need a coordinate system. No coordinate system is shown in the figure.

The green sphere moves with an acceleration of one meter per second squared, and then, the path that the green sphere takes will increase in a certain way.

See Figure 31.

31								
T	0	1	2	3	4	5	6	7
S	0	0,5	2	4,5	8	12,5	18	24,5

In figure 31, a table is shown for the traveled distance depending on the time. The upper horizontal row of the table shows the time since the start of the movement, measured in seconds. The bottom horizontal row of the table shows the distance traveled, measured in meters. The time increases from zero seconds to seven seconds. The road rises from zero meters to twenty-four meters, and fifty centimeters. The path traveled by the green sphere is measured relative to the blue sphere.

The movement of the green sphere is represented graphically as follows.

See Figure 32.

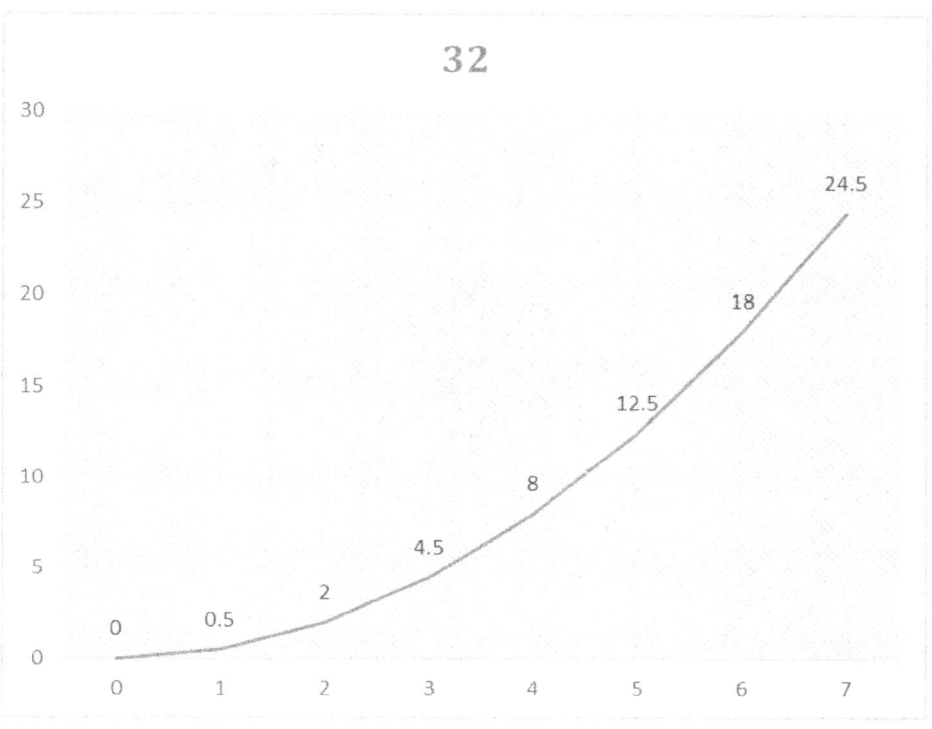

32

In figure 32, the motion graph of the green sphere is shown. The vertical axis of the coordinate system shows the distance traveled. The horizontal axis of the coordinate system shows the instants of time, from zero seconds to seven seconds. It can be seen from the figure that the evil graph starts from zero seconds and ends at the end of the seventh second. Look at the graph.

One second after the green sphere starts, we apply a red thrust to the blue sphere.

See Figure 33.

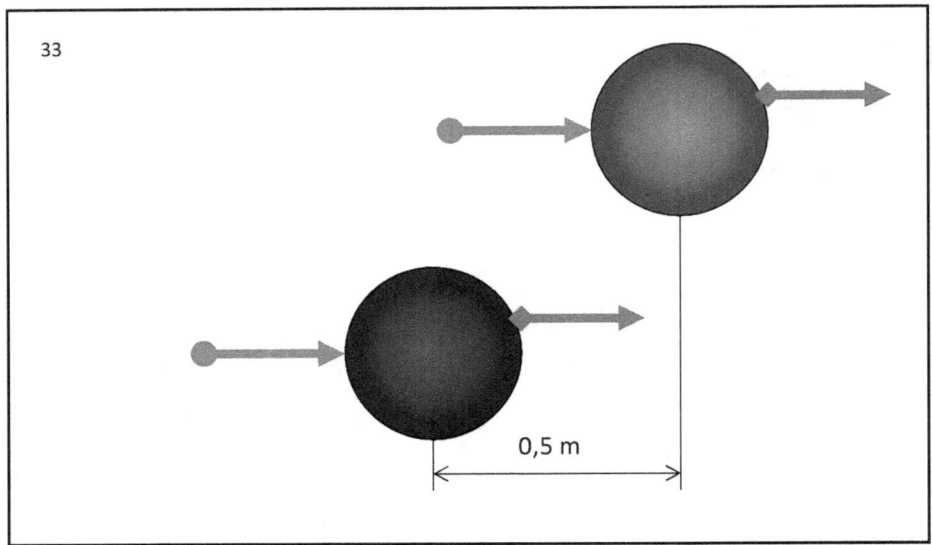

In Figure 33, it is shown that the green sphere continues to have a red push, and that the blue sphere has also already had a red push applied.

The blue sphere starts moving with an acceleration of one meter per second squared. The action of the red thrust on the blue sphere is applied one second after the start of the green sphere. In one second, the green sphere has moved away from the blue sphere by half a meter. This is shown in the figure. The path traveled by the blue sphere in a given time is the same as the path of the blue sphere, but with a delay of one second.

See figure 34.

34								
$T_{n=1\div7}$	1 sec	2 sec	3 sec	4 sec	5 sec	6 sec	7 sec	8 sec
S	0 m	0,5 m	2 m	4,5 m	8 m	12,5	18 m	24,5

Figure 34 shows the blue sphere's motion table. The top row shows the time points, the bottom row shows the traveled distances. The blue sphere moves for seven seconds. Counting seconds starts at **the end of the first second** and ends at the end of the eighth second. I say this because the table shows eight seconds, but the blue sphere is at rest until the end of the first second. From the table it can be seen that in the first second of counting the time, the distance traveled is zero meters. The blue sphere begins its movement at the beginning of the second second, and moves until the end of the eighth second. That's seven seconds. In those seven seconds, the blue sphere travels a distance of twenty-four meters and fifty centimeters. The movement of the blue sphere is represented graphically.

See figure 35.

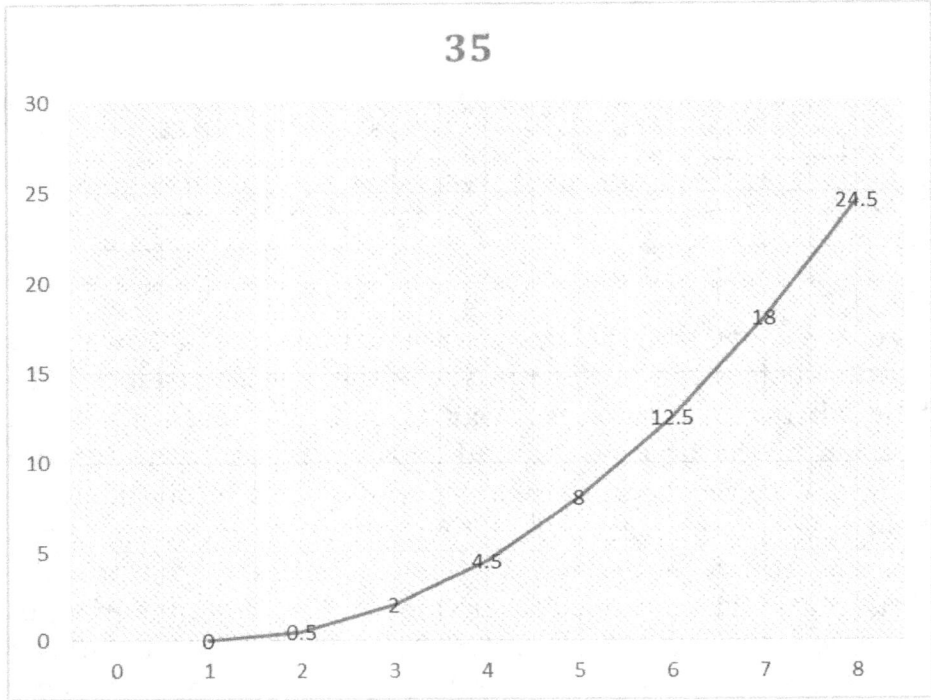

Figure 35 shows that the blue sphere started its movement one second later than the green sphere. The graph shows that the movement of the blue sphere starts at the end of the first second and continues until the end of the eighth second. The blue graph starts at second one and goes up to second eight. Look at the graph.

The movement of the two spheres is represented graphically as follows:

See figure 36.

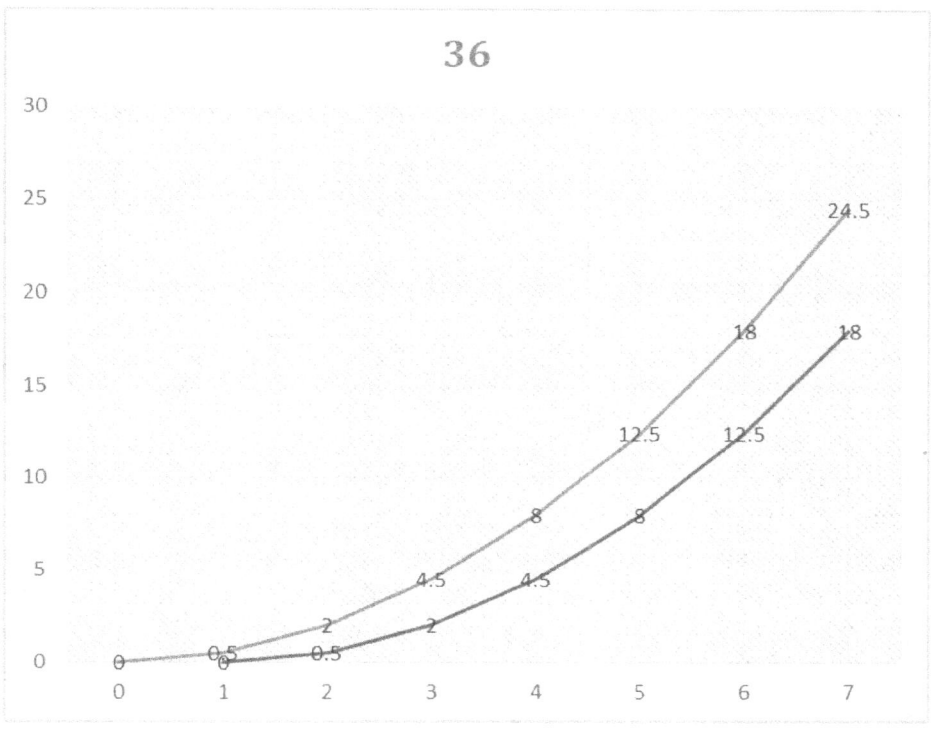

Figure 36 shows graphically the simultaneous movement of the two spheres.

From the graph it can be seen that the green sphere begins its motion at time zero seconds and that the blue sphere begins its motion at time one second.

We will compare the path traveled by the blue sphere with the path traveled by the green sphere.

See figure 37.

37

$T_{n=1\div7}$	0	1	2	3	4	5	6	7
S	0	0,5	2	4,5	8	12,5	18	24,5

	$T_{n=1\div7}$	1	2	3	4	5	6	7
	S	0	0,5	2	4,5	8	12,5	18

In Figure 37, you can see two tables that are placed one above the other. The top table is for the green sphere, the bottom table is for the blue sphere. The tables are placed asymmetrically one above the other. The lower table is shifted to the right, and the distance traveled to the seventh second is shown. The table is shifted because the blue sphere started its motion with acceleration one second later than the green sphere.

We will track how the distance between the two spheres changes.

At the second second after the start of the acceleration motion, the green sphere is two meters from the start of its motion. Look at the red two meters. The second second of the green sphere is the first second of the blue sphere, and it is located at a distance of half a meter from the beginning of the acceleration movement. Look at the red half meter. Therefore, the projection of the distance between the two spheres at the end of the second second from the start of the experiment is equal to two meters minus half a meter, which is one and a half meters.

See figure 38.

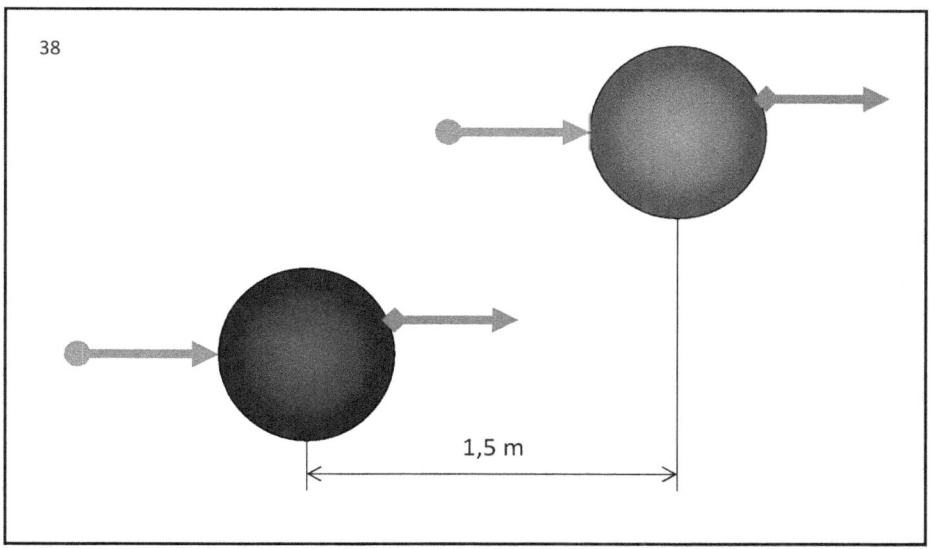

the projection of the distance between the two spheres at the end of the second second is shown . We change the conditions of the experiment. We place the two spheres on a straight line. The direction of the straight line coincides with the direction of motion with acceleration. Thus, the distance projection coincides with the distance.

See figure 39.

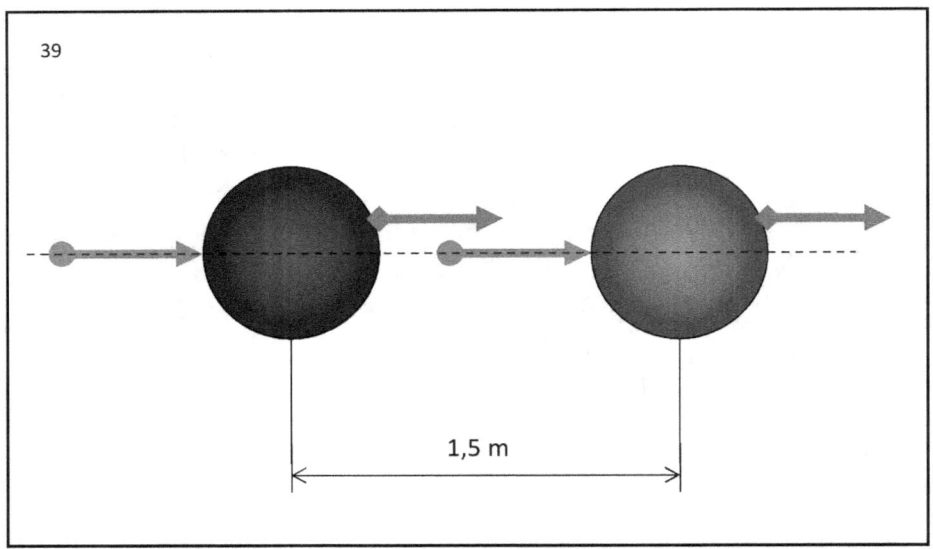

In figure 39, it is shown that the spheres are located in a straight line, and move one after the other. In this way, we directly determine the distance between the two spheres.

The figure shows that at the end of the second second the distance is: (2-0.5=1.5) meters.

At the end of the third second, the distance is: (4.5-2=2.5) meters.

At the end of the fourth second, the distance is: (8-4.5=3.5) meters.

At the end of the fifth second, the distance is: (12.5-8=4.5) meters.

At the end of the sixth second, the distance is: (24.5-18=5.5) meters.

From the calculations we made, it can be seen that the distance between the spheres is constantly increasing, and changes from (1.5) one and a half meters, increases to (2.5) two and a half meters, then (3.5) three and a half , and (4,5)four and a half, and five and a half (5,5).

Every second, the distance between the spheres increases by one meter.

This means that the spheres move **uniformly in a straight line**, relative to each other, at a speed equal to one meter per second.

The results in the table can be presented graphically.

See figure 40.

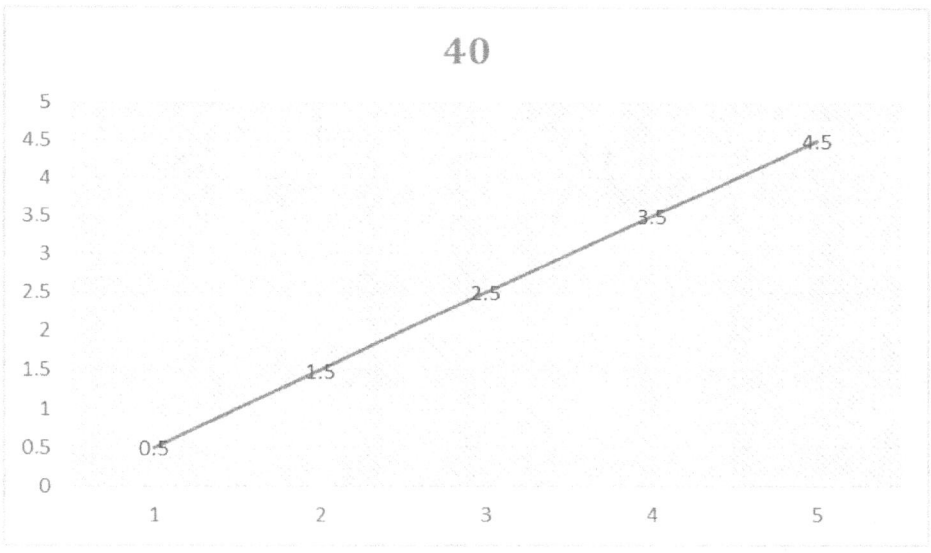

Figure 40 shows how the distance between the blue sphere and the green sphere changes with time.

The graph shows that the two spheres move relative to each other, uniformly and in a straight line at a speed of one meter per second.

Now the question arises: Is it possible to do an experiment that shows some other speed between the two spheres?

The answer is yes, it is possible.

To do this, we change the conditions of the thought experiment we are conducting. We are increasing the delay time of the start of the blue sphere. We apply a force action on the blue sphere, with a delay equal to two seconds, after the start of the green sphere.

See figure 41.

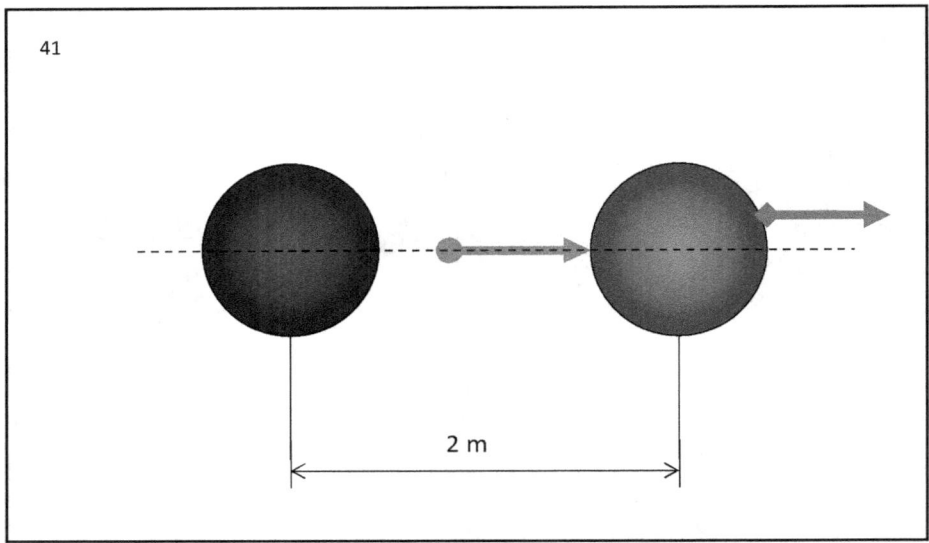

In Figure 41, the blue sphere is shown at rest. A red thrust is applied to the green sphere. The green sphere moves with an acceleration of one meter per second squared. Two seconds after the start, the green sphere will travel a distance of two meters.

See figure above, and see figure below 42.

$T_{n=1 \div 7}$	0 sec	1 sec	2 sec	3 sec	4 sec	5 sec	6 sec	7 sec
S (m)	0 m	0,5 m	2 m	4,5 m	8 m	12,5	18 m	24,5

In figure 42, the table of the distance that the green sphere travels depending on the time is shown. The graph of movement of the green sphere is the same as in the first case.

See figure 43.

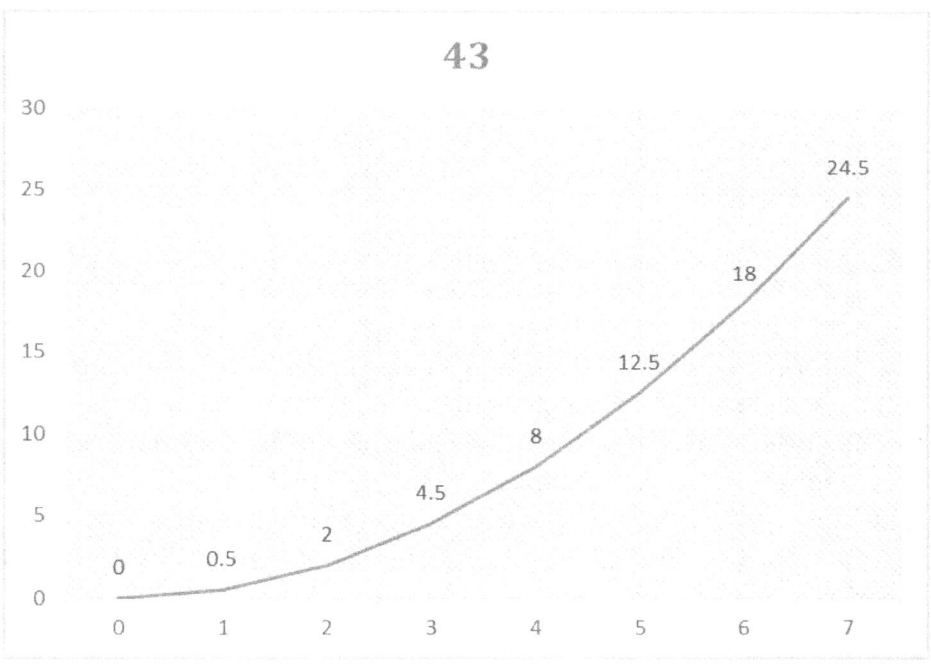

In Figure 43, it can be seen that the green sphere begins its movement at zero seconds and accelerates until the end of the seventh second.

At the end of the second second, from the start of the green sphere's motion, the distance between the spheres is two meters, and then we apply a red thrust to the blue sphere.

See Figure 44.

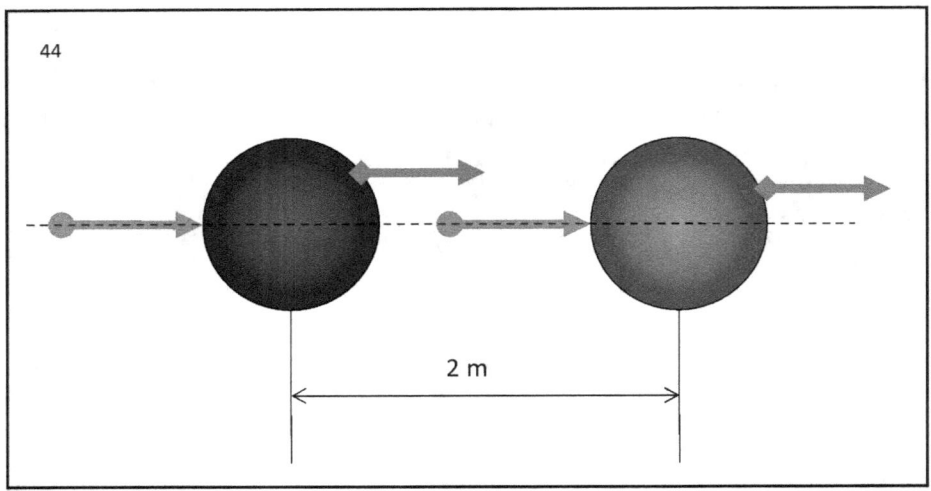

In Figure 44, it can be seen that two seconds after the launch of the green sphere, when the green sphere is two meters from the blue sphere, a red thrust is applied to the blue sphere. The blue sphere moves after the green sphere. The direction of movement of the blue sphere matches the direction of movement of the green sphere. The two spheres are located on a straight line. The blue sphere starts moving with an acceleration of one meter per second squared, but begins its motion at the end of the second second.

See figure 45

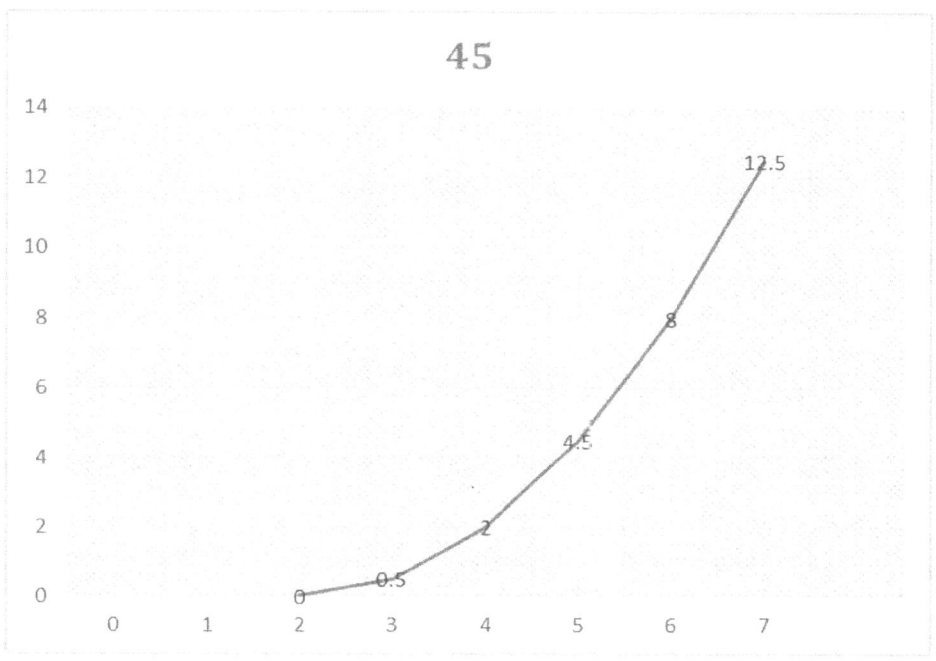

In figure 45, the motion graph of the green sphere is shown. The graph shows that the blue sphere starts its movement at second two, and moves until the end of second seven.

See figure 46.

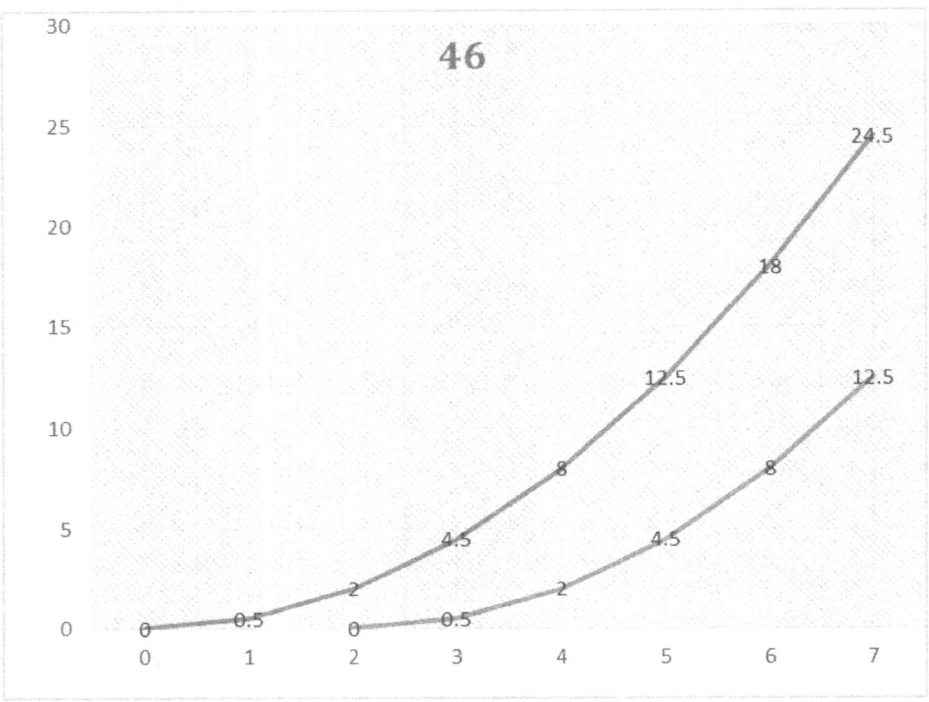

In figure 46, the movement of the two spheres is shown graphically. Blue begins motion with acceleration at second zero, and ends at second seven. Green starts at second two, and ends at second seven.

We compare the path and time tables of the two realms.

See figure 47.

47								
$T_{n=1\div7}$	0 sec	1 sec	2 sec	3 sec	4 sec	5 sec	6 sec	7 sec
S (m)	0 m	0,5 m	2 m	4,5 m	8 m	12,5	18 m	24,5
		$T_{n=1\div7}$	2 sec	3 sec	4 sec	5 sec	6 sec	7 sec
		S (m)	0 m	0,5 m	2 m	4,5 m	8 m	12,5

In Figure 47, two tables are shown. The above table is on the green sphere. The bottom of the blue sphere. The tables are shifted in such a way that the road and time results on the green sphere are compared with the results on the blue sphere.

The distance between the two spheres increases as follows:

At the end of the second second, the distance is (2-0=2) two meters.

At the end of the third second, the distance is (4.5-0.5=4) four meters

At the end of the fourth second, the distance is (8-2=6) six meters.

At the end of the fifth second, the distance is (12.5-4.5=8) eight meters.

At the end of the sixth second, the distance is (18-8=10) ten meters.

At the end of the seventh second, the distance is (24.5-12.5=12) twelve meters.

Each successive kunda, the distance between the two spheres increases by two meters. This means that the two spheres are moving relative to each other at a speed of two meters per second.

See figure 48.

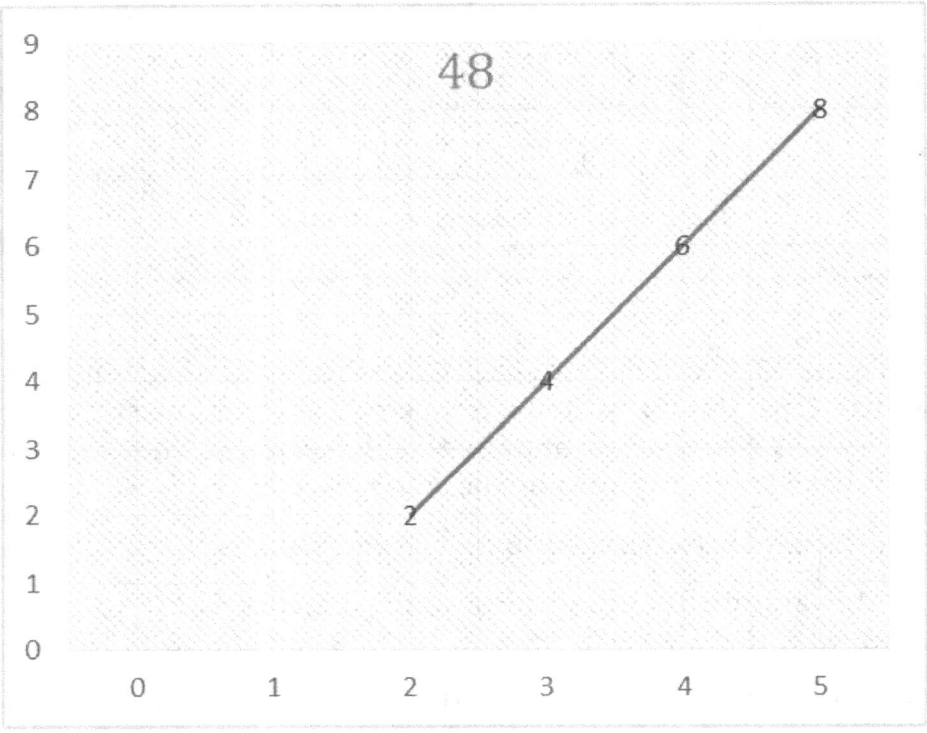

In figure 48, uniform rectilinear motion of the two spheres relative to each other is shown. The green sphere is moving relative to the blue one at a speed of two meters per second.

The movement starts at second two and ends at second seven.

We have done experiments which show that we are in a position to obtain different relative velocities between the two spheres. This result allows us to deduce a natural law that states that:

Uniform rectilinear motion between two physical bodies can always be represented as motion with acceleration of these two

physical bodies.

This means that any **relative motion** can be represented by **absolute motion** with acceleration.

From a philosophical point of view, the last judgment is strange, and needs further analysis, and relevant conclusions and conclusions. The conclusions drawn will contribute to the enrichment of some of the philosophical categories. This will be done at a later stage in the research process we are doing.

11. SENSATION OF THE ACTION OF FORCE.

In the reality surrounding us, there is another fact to which we have to pay special attention. We are talking about the phenomenon of "sensation of acceleration" and "sensation of force action", which can be combined into one, a phenomenon designated as "sensation of force action and movement with acceleration". This is a part of every person's everyday life, that is why it is always clear to everyone that when the train starts, the passengers in it "feel" this by the push they receive at the first moment and the force acting afterwards, which has the opposite direction of the direction of travel. In this case, no one is surprised that the backs of the seated passengers are pressed against the backrests of the train.

The reason for this phenomenon is the inertial force, which is sometimes called the fictitious force.

Everything said so far is in agreement with Newton's third law, which states that for every action there is an equal and opposite reaction.

To these considerations we must add Newton's second law, from which it is clear that when a body that has some mass a force acts, the body begins to move with acceleration .

And indeed, train passengers immediately understand, with a glance out the window, that they are moving at an increasing speed, which is constant acceleration.

We deliberately separate "sensation of force action and motion with acceleration" into an independent phenomenon with its own essence that we must understand.

The question arises, what is the cause of the phenomenon "sensation of force action and movement with acceleration"? The answer to the question we give is that the phenomenon of "sensation of force action and movement with acceleration" is the result of the **complex action of Newton's second and third laws**.

Now consider an elevator that has passengers in it, and unfortunately at some point in time, the rope breaks. See figure 49.

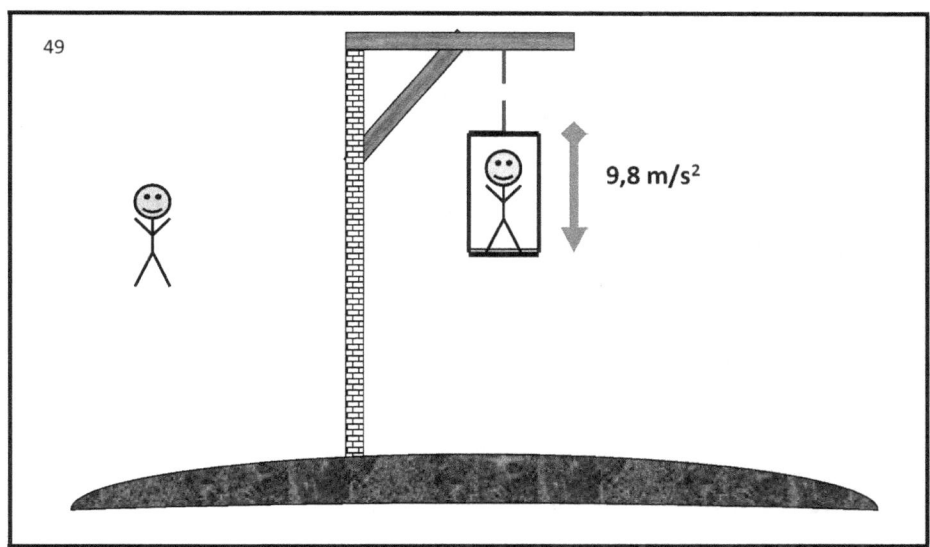

In figure 49, a portion of the earth's surface is shown, a strong vertical support on which a horizontal beam is fixed. The elevator is roped to the girder. The rope is broken. For our consideration, it is not important whether the elevator was in motion or at rest at the time the rope broke. What is important is that the elevator will begin to fall toward the earth's surface, and it will move at an acceleration of nine whole eight tenths of a meter per second squared. The reason for this fall with acceleration is that the elevator, and the passengers in it, are in the gravitational field of the Earth, and experience the action of the force of the Earth's gravitational attraction.

The quantitative characteristic of this force was shown by Newton, and is known as the law of gravitational attraction:

The force of gravitational attraction between two bodies is equal to the mass of the first body times the mass of the second body divided by the distance between them squared.

Passengers in the elevator have no "sensation of the action of the force of the gravitational pull of the Earth." On the contrary, they will be convinced that they are at rest or in uniform rectilinear motion, and are not acted upon by forces which cause acceleration. Passengers in the elevator are convinced that their state is determined in accordance with Newton's first law:

When no force acts on a body, it is in a state of rest or uniform rectilinear motion .

It should be noted that similar thought experiments with elevators were conducted by Einstein to clarify the nature of inertial and non-inertial reference frames. These thought experiments are extremely important, and through proper analysis, can reveal fundamental relationships between motion, rest, relative, absolute.

At the beginning of our presentation, we defined a clear dependence confirmed in practice:

Always, and only, the simultaneous, complex action of Newton's second and third laws is the cause of the phenomenon "sensation of the action of force and movement with

acceleration".

We have reason to conclude that for passengers in the elevator, the complex effect of Newton's second and third laws is not valid.

Newton's second and third laws are at the foundation of physics. These two laws are of fundamental universality, and necessarily encompass all possible phenomena in the One Infinite Reality. The simultaneous operation of the second and third laws show the essence of absolute motions in the One Infinite Reality. There are no exceptions.

It is necessary to find out and indicate the reasons why passengers in the elevator do not have a "sensation of the action of force and movement with acceleration".

See figure 50.

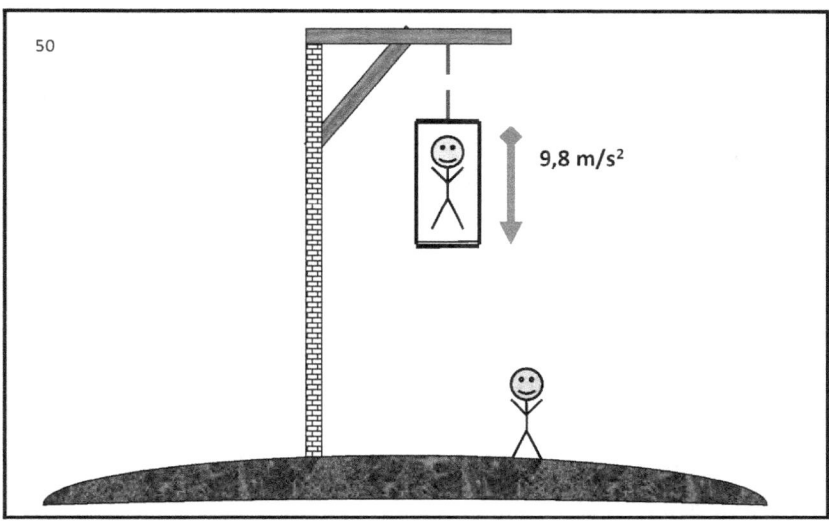

Figure 50 shows the supporting frame, the broken rope, the elevator and a passenger in it. The elevator falls to Earth. The elevator has no windows and the passenger cannot understand what is happening to him. The passenger feels that he is in a state of weightlessness. The traveler concludes that he is in deep space and his state is described by Newton's first law. The passenger is convinced that there is no force acting on the elevator, and the elevator is at rest, the elevator is in a state of weightlessness.

There is a second person on Earth watching the falling elevator.

A telephone connection exists between the passenger and the observer.

The observer calls on the phone and tells the passenger that he is falling and when he hits the ground, he will most likely die. The traveler replies that this is not true and that he is in a state of weightlessness and that he is at rest and that the observer is making some mistake.

The observer replies that there is no mistake, that he is firmly planted on the earth's surface, that he feels his weight, and that he watches the elevator fall.

The passenger smiles and says if you really feel weight it's because you're moving towards me with acceleration. You are hallucinating or dreaming. This is the truth.

See figure 51.

EINSTEIN'S THIRD MISTAKE

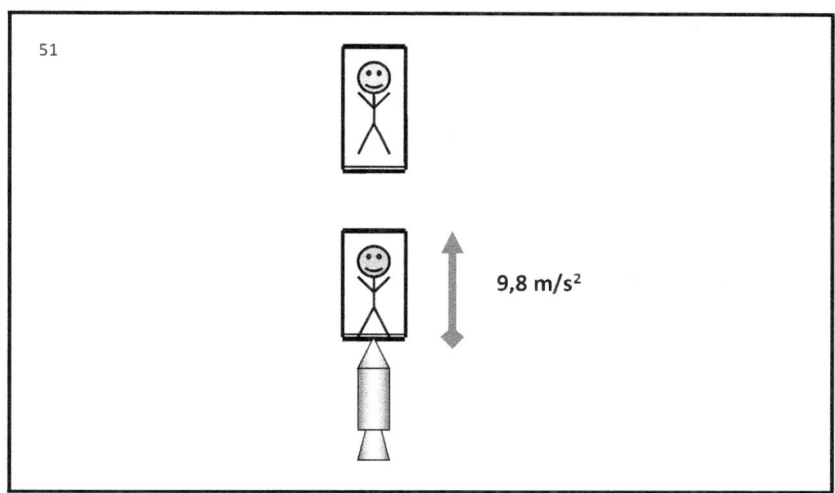

51

Figure 51 shows the passenger in the elevator, the observer who is in a second elevator. A rocket is placed at the bottom of the second elevator, which pushes the elevator with the observer up. The elevator with the observer is moving with an acceleration of nine whole and eight tenths of a meter, per second squared.

The passenger in the upper elevator calls the observer and asks him what he is doing now.

The observer answers that he is in an elevator that is moving with upward acceleration.

The passenger asks him what he feels.

The observer says that he has landed firmly on the bottom of the elevator, and he feels the action of force and motion with acceleration, in the same way as when he landed on the earth's surface.

The passenger in the upper elevator replies that this is the true state of motion and that this is no longer a dream.

The observer asks why this is the true state.

The passenger replies that he is sure because there is a

principle that says:

Always, and only, the simultaneous, complex action of Newton's second and third laws is the cause of the phenomenon "sensation of the action of force and movement with acceleration".

The principle thus defined shows the difference between relative and absolute motions which take place in the One Infinite Reality.

This principle shows that the force defined in Newton's second law is fundamentally different from the force of gravitational attraction between bodies.

12. STRENGTH. APPLICATION POINT OF ACTION.

Newton's second law states that the force acting on a body is equal to the product of the acceleration and the mass of the body moving with the acceleration.

In this case, the acting force, vingi, has an applied point of action. A site of action is a specific location on the body. The place of action is a surface on which at least two bodies are pressed against each other. This surface in physics is called an application point. From a philosophical point of view, the concept of a point, by which the phenomenon of a point is denoted, is subject to serious criticism. The problem is that there is no point phenomenon in the One Infinite Reality. The concept of a point serves only to denote a human abstraction, in the mind of man. In the science of mathematics, the concept of a point is used, and it has a certain mathematical content, which is again an abstraction. In physical science, the concept of point should be replaced by the concept of place.

This is how Newton acted in "Mathematical Principles of Physics". In the "Principles", Newton did not use the concept of point. In the "Principles", Newton defines the phenomenon of place, and uses the concept of **place** whenever he should use the concept of point.

This fact is extremely important for the research we are doing and should be remembered.

13. TYPES OF FORCES. MANIFESTATION OF POWER. CAUSE EFFECT.

There are two types of forces in modern physics. Real forces, and fictitious forces. Fictitious forces appear and act when there is **simultaneous mutual action** between at least two things.

Simultaneous mutual actions are denoted by the term

ВЗАИМНОДЕЙСТВИЕ

.

The word

ВЗАИМНОДЕЙСТВИЕ

, is written in Slavic-Bulgarian Cyrillic.

I suggest, in English writing, to use the word

MUTUALISACTION .

I hope that specialists in this field will accept my suggestion, and when necessary, will cite its origin.

The word

ВЗАИМНОДЕЙСТВИЕ

= *MUTUALISACTION* , is a

verb, and means parallel, simultaneous actions performed by **whole** things. The concept of **interaction** = *ВЗАИМНОДЕЙСТВИЕ* = *MUTUALISACTION*, is a philosophical category. Through the category **interaction** = *MUTUALISACTION*, the mutual action between two whole things is indicated. Each of the two wholes interacting with each other is always a **whole part** of **the whole** One Infinite Reality.

An entire part of the One Infinite Reality is defined by the absolute motion that that part performs in relation to the whole One Infinite Reality.

Fictitious forces appear, and act, when some absolute motion is related to another absolute motion. Typical examples of this are the way they appear, the Coriolis force, the Cup Force and the way quantum mechanical objects interact with each other.

The Coriolis force occurs when the absolute rotational motion of the planet Earth is related to the absolute motion of the Foucault pendulum.

The force of the cup occurs when the absolute rotational movement of the cup about some center is related to the rotational movement of the platform about its own center.

The force of rotation, on the back of the cup, appears when the absolute rotational movement of **the entire** cup, around some axis, relates to the absolute rotational movement of **the entire** arrow, indicating the direction of the centrifugal force, around the same axis.

Note: The last two judgments are explained in the post Dark Energy Dark Matter.

Typical cases of **interactions** = *MUTUALISACTION*, take place between quantum mechanical objects. The science of quantum mechanics studies and describes how one whole quantum relates to another whole quantum through the phenomenon of *MUTUALISACTION*.

In this way, the quantum becomes **whole** in time and **whole** in space. Thus, the quantum can perform *MUTUALISACTION*, and change **quantum**, in portions, which is **a change of state**. Thus, every **quantum**, change of **state**, is a multiple of Planck's quantum, the constant h.

The change of **state** of **the quantum** involves all **parts** of **the whole** quantum, whereby **the whole** quantum interacts with **the whole One Infinite Reality**, the **whole** with **the whole**.

The change of state takes place in **the present** and is logically absolutely simultaneous for **all**, One, Infinite, Reality.

In this sense, the moment of the present is a time interval equal to zero, and separates the past from the future.

The absolute present is relative, only and only, generally **to** the past, and only, and only, generally **to** the future. In this way, the parallel changes of reality appear. And this, again, is **a change of states** , through interactions=

MUTUALISACTION

The parallel changes themselves receive being in the only present, where and in which it is possible to relate to one another, whole things to other whole things. These are relations of some **whole parts** to other **whole parts** . Whole parts can be different **whole parts** of a **whole** , or different **whole parts** of different **whole** things.

The change of states is a process that proves the existence of logically absolute simultaneity, and in this connection the extremely important question arises:

What is the carrier of this simultaneity, or to put it another way, what is the phenomenon through which this simultaneity can be transformed, reduced to a quantifiable physical quantity?

The answer to these two questions boils down to finding physical evidence, empirical data and facts showing unequivocally the existence of the carrier of parallel motions, which in modern science are known as action at a distance, in classical Newtonian mechanics, or as non-local interaction, in quantum mechanics, or as motion with an infinitely high speed, in the theory of relativity, which in our hypothesis is **a change of states, through interaction** =

MUTUALISACTION

Once again we have to pay attention to the fact that modern science is unable to indicate the carrier of a change of states, through

MUTUALISACTION

interaction, or what is the same, to indicate some new field that makes possible the non-local *MUTUALISACTION* = **interaction** , between things.

In this regard, and as a result of the analysis, we propose that the carrier of the distant action be called, denoted by the term **field of effort** .

In modern physics there is the idea that distant action is motion at an infinitely high speed. In the book "Einstein's Second Mistake", I explained and proved that the expression " **motion with infinitely great speed** " is incorrect. What human science calls " **motion with infinitely great speed** " **is not speed** .

But this does not mean that such a phenomenon does not exist. What people call " **motion at infinite speed** " is **a change of states** , and is a fundamental property of **the One Infinite Reality** .

It is precisely this process by which **the change of states takes place** that I call **reciprocity**= *ВЗАИМНОДЕЙСТВИЕ* = *MUTUALISACTION* .

14. PRINCIPLE OF UNIFORMITY.

In the hypothesis I present, Einstein's **Principle of Equivalence** is replaced by **the Principle of Equality** . This means that the movement of a body that falls in a gravitational field is **uniformly rectilinear** , or is in a state of **relative rest** .

See figure 52.

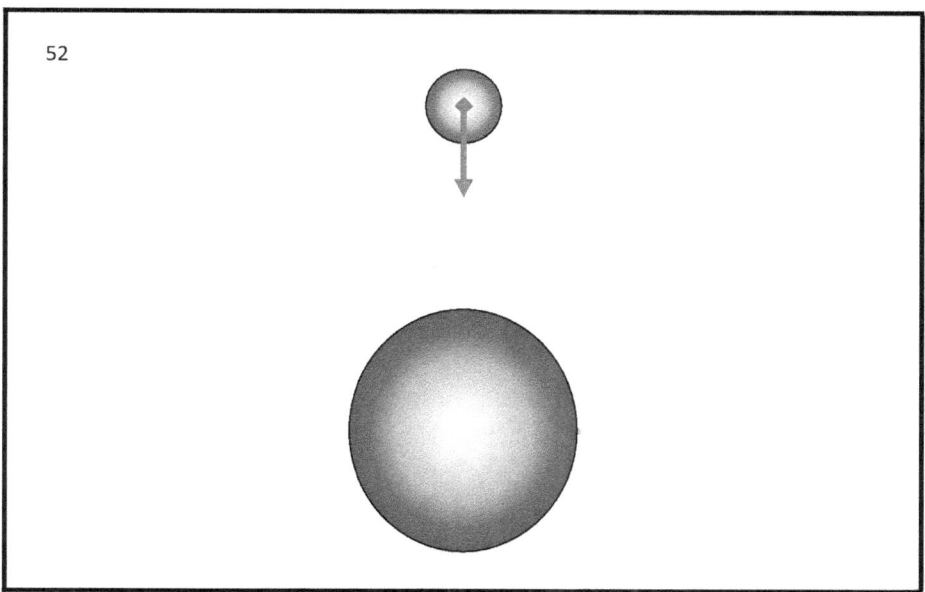

In Figure 52, two spheres are shown. The large sphere is stationary, and possesses a large mass and a powerful gravitational field. The small sphere "falls" towards the large sphere, and moves with **acceleration** , but does not feel the action of a force and does not feel that it is moving with **acceleration** . This is Einstein's **Equivalence Principle** .

We replace Einstein's **Principle of Equivalence** with **the Principle of Equality**.

See Figure 53.

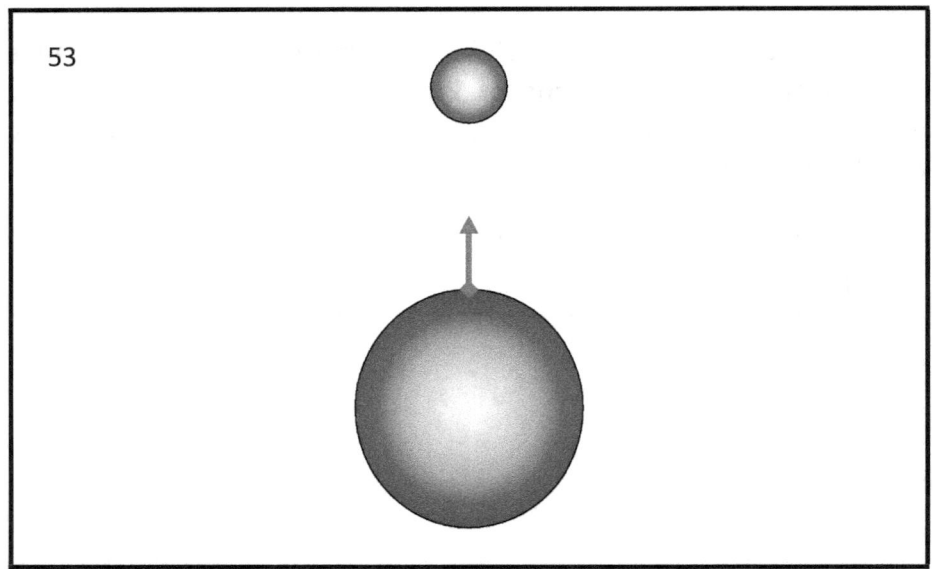

In Figure 53, two spheres are shown. The large sphere is stationary, and possesses a large mass and a powerful gravitational field. The small sphere does not feel "action of force", and it does not feel "motion with acceleration", therefore the small sphere is in **a state of rest or uniform rectilinear motion**. This means that the surface of the large sphere is moving with **acceleration** towards the small sphere. It is necessary to emphasize that only and only **the surface** of the large sphere moves with **acceleration** towards the small sphere. The center of the large sphere is stationary relative to the small sphere. From what I have said, it follows that the great sphere **is constantly increasing its radius**, and the entire surface of the great sphere is **moving away** from the center of the great sphere, with **an acceleration of** . To put it short and simple, the large sphere inflates like a balloon.

I know very well that some of the readers will strongly object that this is impossible.

I continue to maintain that this is possible and that:

The "BORDER" of the entire One Infinite Reality, moves away from each whole part of it with increasing acceleration, and variable acceleration.

The necessary and sufficient condition for continuous motion with increasing acceleration and variable acceleration is that the One Infinite Reality must be **infinite**. I have to recall that at the beginning of the exhibition, we created a definition area.

In the definitional realm, principle four states: Reality is **infinite**.

15. GRAPHIC REPRESENTATION

The One Infinite Reality is "expanding" with increasing acceleration. The incremental acceleration is a constant total, integral **acceleration** . In specific places, on the One Infinite Reality, the local acceleration is different. The local acceleration can be differentially decreasing, differentially increasing, or differentially constant. The One Infinite Reality is spatially three-dimensional. The acceleration of the Spatially three-dimensional One Infinite Reality takes place absolutely simultaneously along the three spatial dimensions. The three spatial dimensions are presented to human thinking through a three-dimensional coordinate system.

See figure 54.

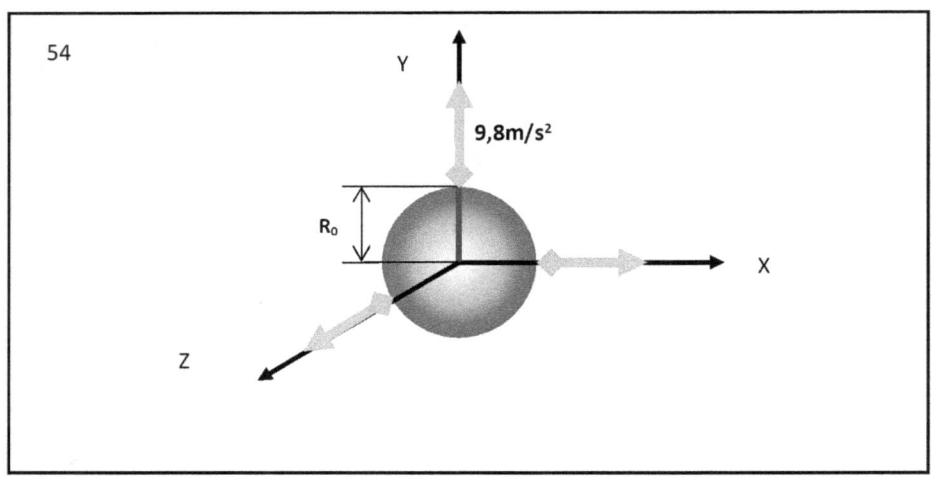

In figure 54, a coordinate system is shown which consists of three axes. The origin of the coordinate system is located at the center of a sphere.

The coordinate system and the sphere are located at the center of the One Infinite Reality. We assume that the sphere is the planet Earth. The acceleration of the Earth's surface, relative to the center of the planet Earth, is equal to nine whole eight tenths of a meter per second squared. Acceleration is shown in green arrow, radius is shown in blue. This means that the length of the radius of the planet Earth increases with an acceleration equal to nine whole and eight tenths of a meter per second raised to the second power. This means that after some time, the size of the planet Earth will be twice as large.

See figure 55.

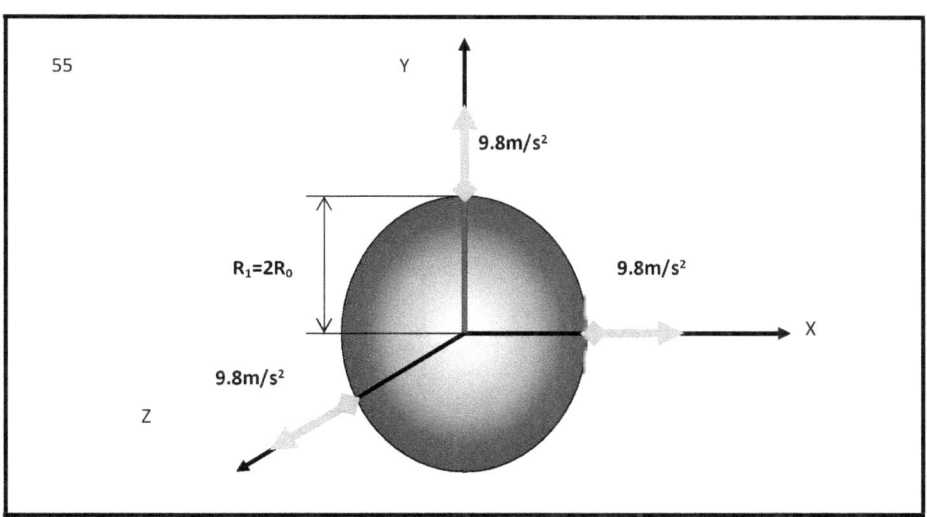

In figure 55, the coordinate system and the planet Earth are shown. The radius of planet Earth is twice as large.

The intelligent, thinking, human beings that inhabit planet Earth do not notice the increase in the size of the Earth. The reason for this is that all solid bodies and objects that are on the surface of the Earth increase in size in proportion to the increase in the radius of the planet Earth. When the magnification is proportional, then the ratio of the spatial dimensions of the different objects does not change. The ratio is kept constant. The ratio is a constant.

When the ratio of spatial dimensions is a constant, then the increase in spatial dimensions cannot be registered by measuring instruments. It cannot be noticed by the researchers who measure the distances.

See figure 56.

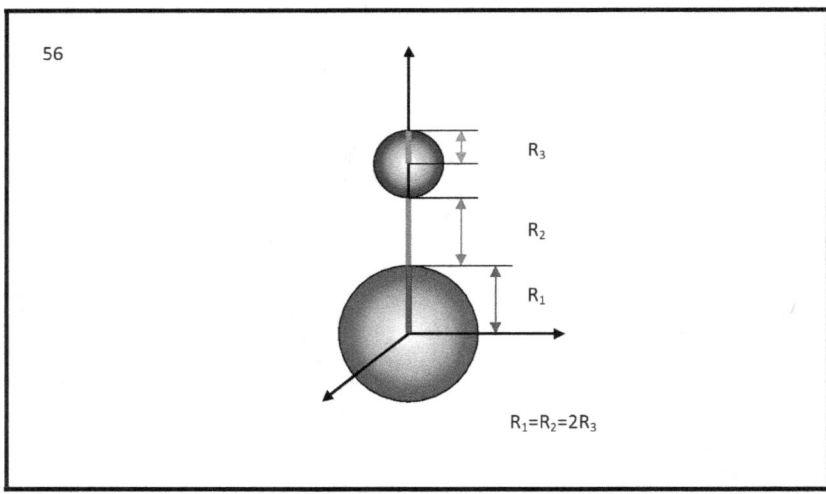

In figure 56, the coordinate system and two spheres are shown. A large sphere and a small sphere. The large sphere is the planet Earth before it increased its radius. The radius of planet Earth is shown in blue. The small sphere is located on the vertical axis of the coordinate system. The radius of the small sphere is shown in red. The radius of the planet Earth is twice the radius of the small sphere. The distance between Earth and the small sphere is shown in green. The distance between the Earth and the small sphere is equal to the radius of the Earth. The distance between the Earth and the small sphere does not change. The earth and the small sphere are at rest relative to each other.

The radius of the earth is doubled, by an acceleration of nine whole and eight tenths of a meter, per second squared.

See figure 57.

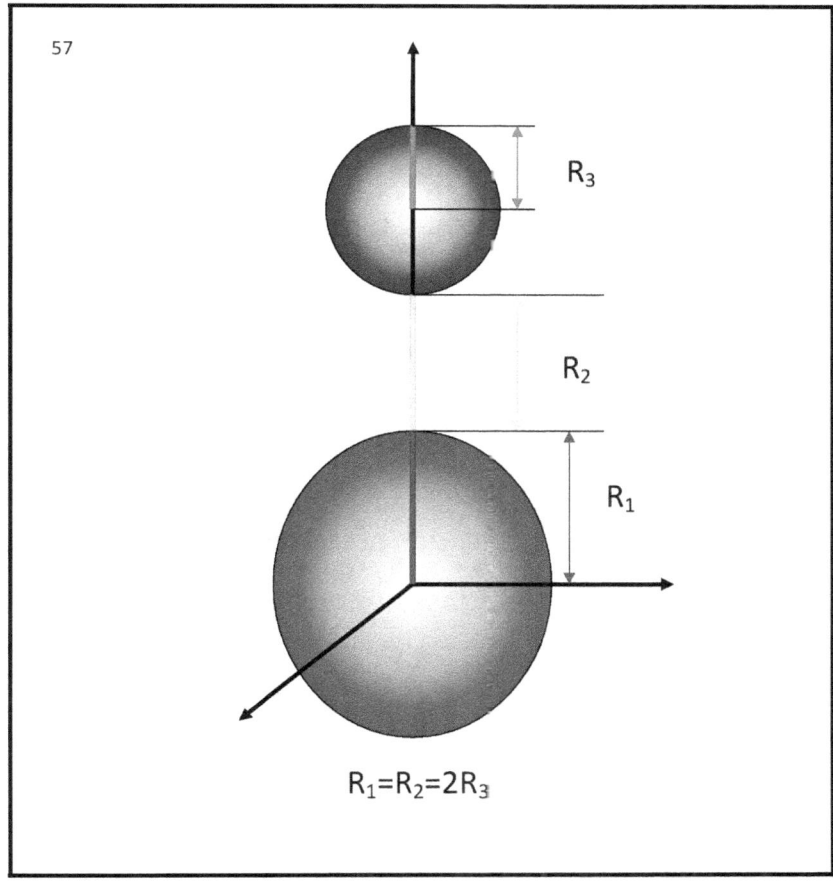

In figure 57, are shown the planet Earth, the small sphere coordinate system.

The radius of the Earth has doubled.

The radius of the small sphere has doubled.

The distance between the Earth and the small sphere has entrained twice.

Under these conditions, the relations between the dimensions are kept constant.

The ratio between the radius of the Earth and the radius of the small sphere does not change.

The ratio between the radius of the Earth and the distance to the small sphere does not change.

The ratio between the radius of the small sphere and the distance does not change either.

All physical bodies that exist on planet Earth have increased their spatial dimensions, and are now twice as large. The researcher who will perform the measurement is twice as big. The explorer's meter is twice as large.

The magnification of the Earth, the magnification of the small sphere, the magnification of the distance, are not noticeable.

The result of the measurement is that the two spheres retain their dimensions, and the two spheres are at rest relative to each other.

16. CONDITION OF RELATIVE REST

The radius of the Earth is a certain length. The surface of the Earth is moving away from the center of the Earth at an acceleration of nine whole eight tenths per second squared. The radius of the small sphere is twice the radius of the Earth. The dimensions of these two radii are relative to each other at rest. Therefore, the acceleration with which the radius of the small a sphere increases is twice as small as the acceleration of the Earth. The acceleration of the radius of the small sphere is equal to four whole and nine-tenths meters per second squared. The number four whole and nine tenths is half of the number nine whole and eight tenths.

See figure 58.

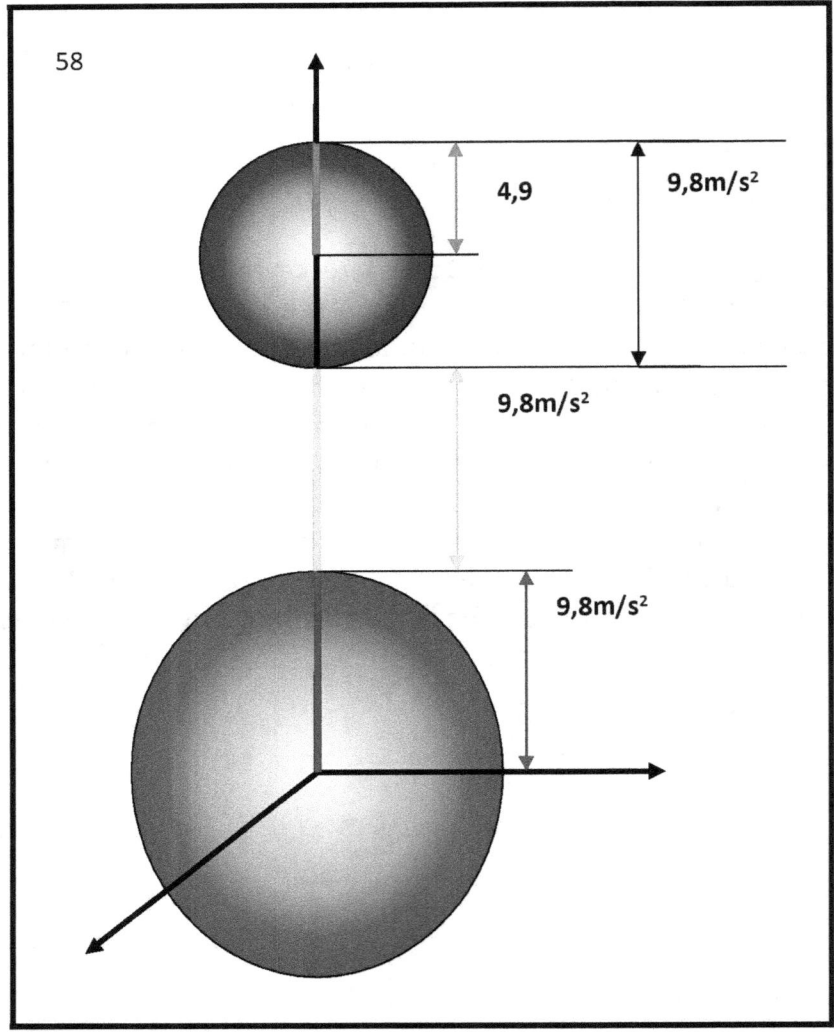

In Figure 58, the Earth, the small sphere, and the distance between the Earth and the small sphere are shown. Shown are the accelerations with which the sizes of the two radii increase, and the acceleration with which the distance between the Earth and the small sphere increases. At these accelerations and at these distances, the Earth and the small sphere are in a state of relative rest.

The state of relative rest is also possible at other distances between the Earth and the small sphere.

See figure 59.

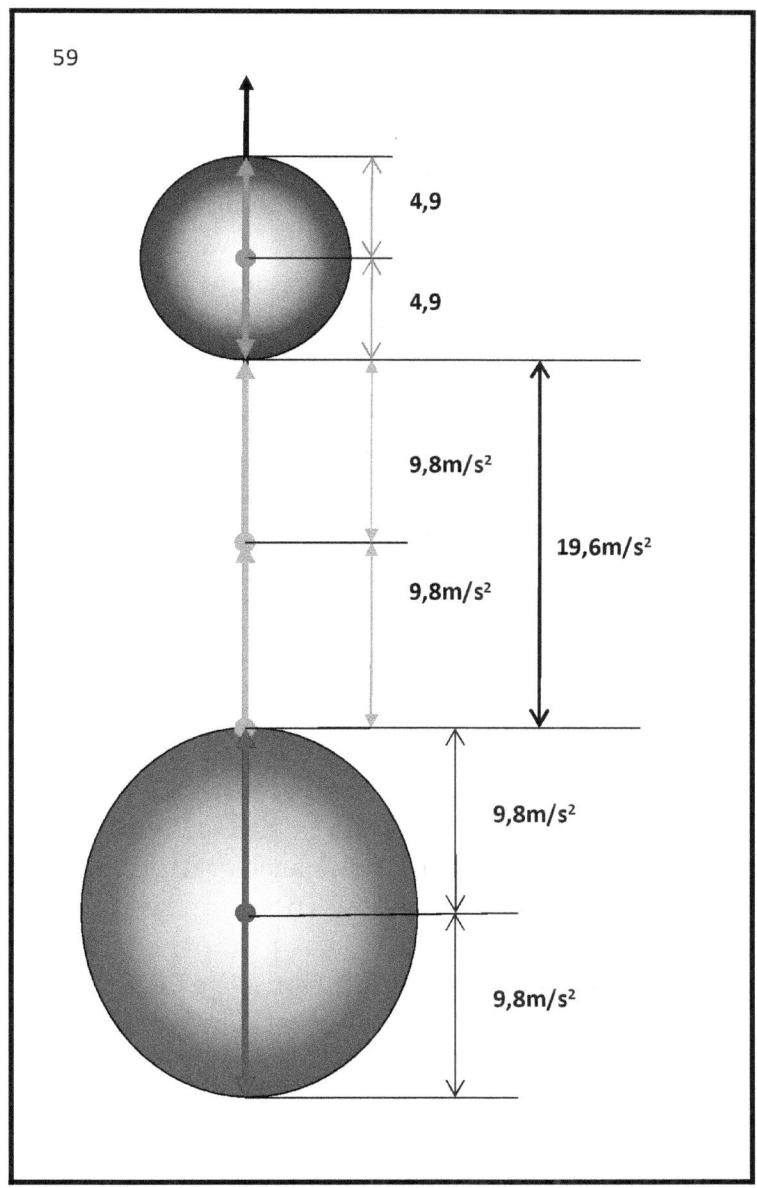

In Figure 59, a large sphere-Earth, a small sphere, and **the vertical** axis of the coordinate system are shown. The vertical axis of the coordinate system starts from the center of the Earth and ends above the surface of the small sphere. This is the black arrow visible at the top.

Shown is the diameter of the Earth, which is blue, and the acceleration of the Earth's surface relative to the center of the Earth. These are two blue radii that start from the center of the Earth, and are perpendicular. One at the top, the other at the bottom. On the right are numbers and double arrows that show the magnitude of the ground acceleration. Nine whole and eight-tenths meters per second squared is the earth's acceleration, relative to the center of the earth.

Shown is the diameter of the small sphere, in red, and the accelerations of the radii of the small sphere, in red. The accelerations of the two radii of the small sphere are shown with red double arrows, numbers. The accelerations are in opposite directions, from the center of the small sphere to the surface of the small sphere. The acceleration of the surface of the small sphere, relative to the center of the small sphere, is equal to four whole and nine-tenths meters per second squared.

The distance between the Earth and the small sphere is shown, which is twice as large compared to the distance in the previous figure. The long distance is shown with a green line. The magnitude and direction of the acceleration is indicated by a green arrow. The numbers show the numerical values of the accelerations. Twice the distance, has twice the acceleration. At these dimensions and these accelerations, the Earth and the small sphere are again in a state of relative rest relative to each other.

The figures show that absolute motions with acceleration are relative to each other and are at relative rest.

The figures show that relative rest is a special case of

absolute motion with acceleration.

This means that any **relative rest can be reduced to absolute motion with acceleration.**

I will emphasize once again that this is an extremely important, fundamental property of rest and motion, and that modern physics has not paid enough attention to this fact.

The condition for relative rest is:

$$\frac{a_n}{S_n} = const.$$

Where:

$$n = 1; 2; 3; \ldots \to \infty$$, is a sequence number.

a_n - is the acceleration with an ordinal number that corresponds to a precisely defined distance S_n having the same ordinal number.

S_n - is a distance with an ordinal number that corresponds to a well-defined acceleration a_n, with the same ordinal number.

$const.$ - is a numerical constant that is the same for the entire set consisting of relations between accelerations and distances that have the same ordinal number.

17. THREE-DIMENSIONAL REALITY. ONE DIMENSIONAL REALITY.

The One Infinite Reality is three-dimensional. From the point of view of the science of mathematics, the One Infinite Reality can be represented by more than three dimensions. At this point, it's redundant.

A three-dimensional space is represented by a three-axis coordinate system. A three-dimensional space that is in a state of acceleration relative to its center increases in size along the three axes.

Increasing the size of the three axes of the coordinate system is absolutely simultaneous.

The increase in the size of the three axes of the coordinate system is carried out with the same acceleration.

See figure 60.

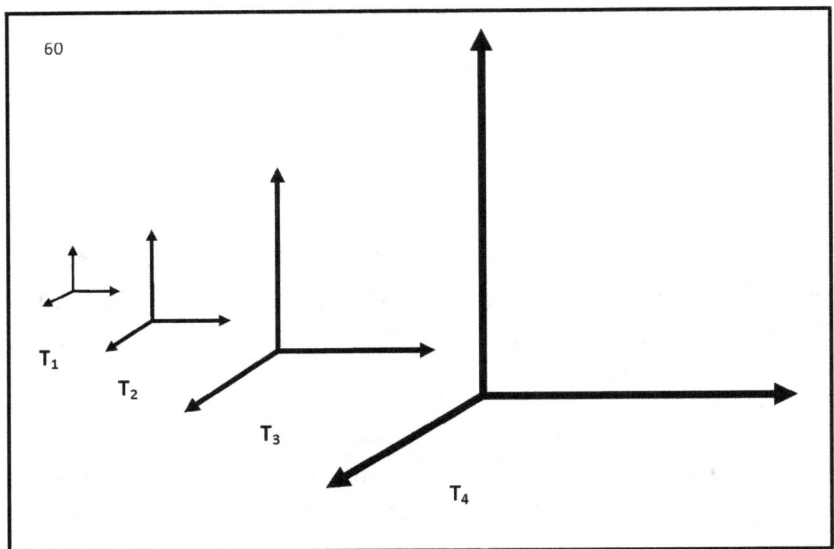

60

In figure 60, four coordinate systems are shown which have different dimensions.

It is a coordinate system that scales the size of the three axes in four instants of time. At each subsequent moment of time, the coordinate system is twice as large as the previous one. Each of the four coordinate systems, at any given moment in time, is at rest relative to itself.

Each one of the axes of the three-dimensional coordinate system represents a One-Dimensional Reality.

See figure 61.

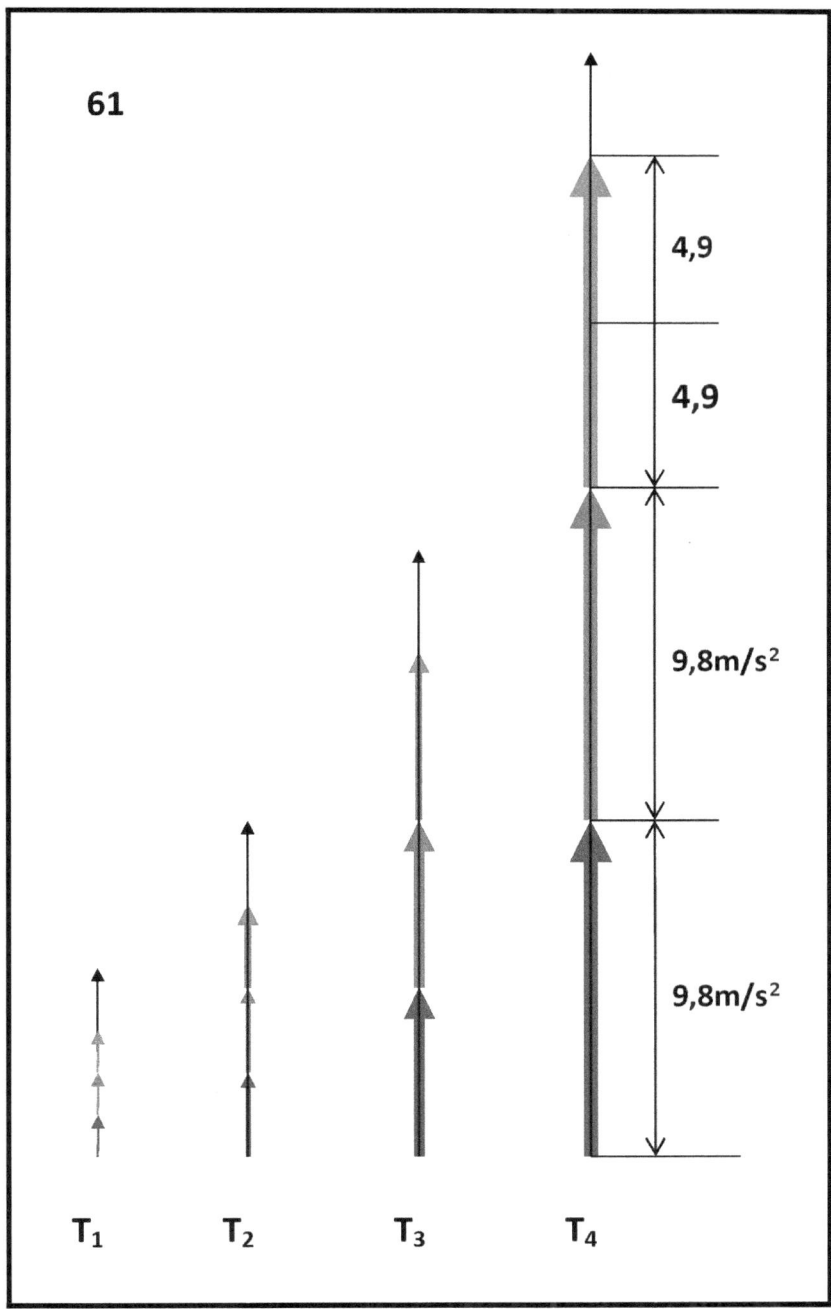

In Figure 61, only the vertical axis of the three-dimensional coordinate system is shown. The vertical axis is a one-dimensional reality. Four consecutive moments of time, of one-dimensional reality, are shown. Accelerations and distance increments are shown. In blue, the acceleration and increase in the size of the radius of the planet Earth is shown. The green color shows the acceleration and increase in the size of the distance between planet Earth and the small sphere. In red, the acceleration and increase in size of the diameter of the small sphere is shown.

The thin black arrow is the vertical axis of three-dimensional reality.

The growth of the distances, depending on the growth of time, is presented graphically.

See figure 62.

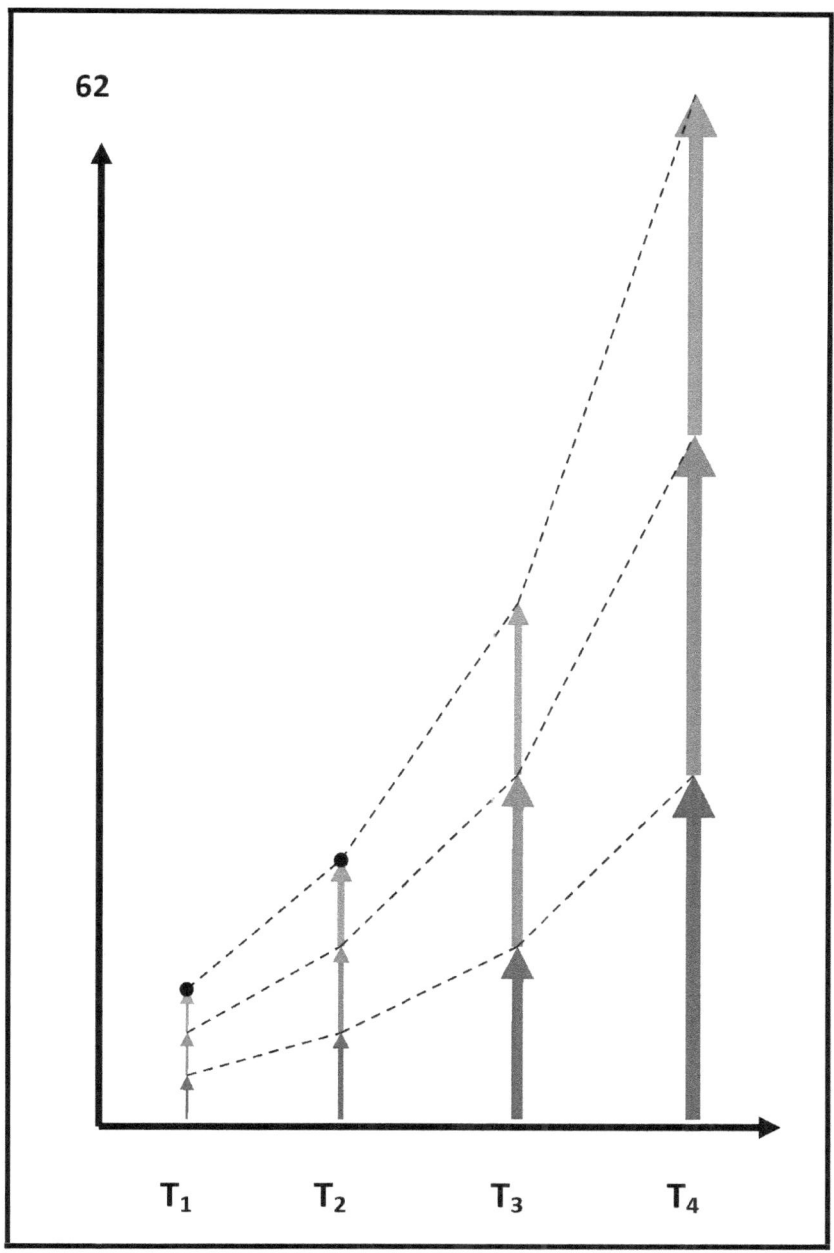

In figure 62, the graph of the relationship between

increasing distances and increasing time is shown. Four distances are shown, at four consecutive points in time.

The following graph shows a one-dimensional reality that has **an increasing acceleration coefficient** equal to one meter per second squared. The time of existence of one-dimensional reality is equal to four seconds.

See figure 63.

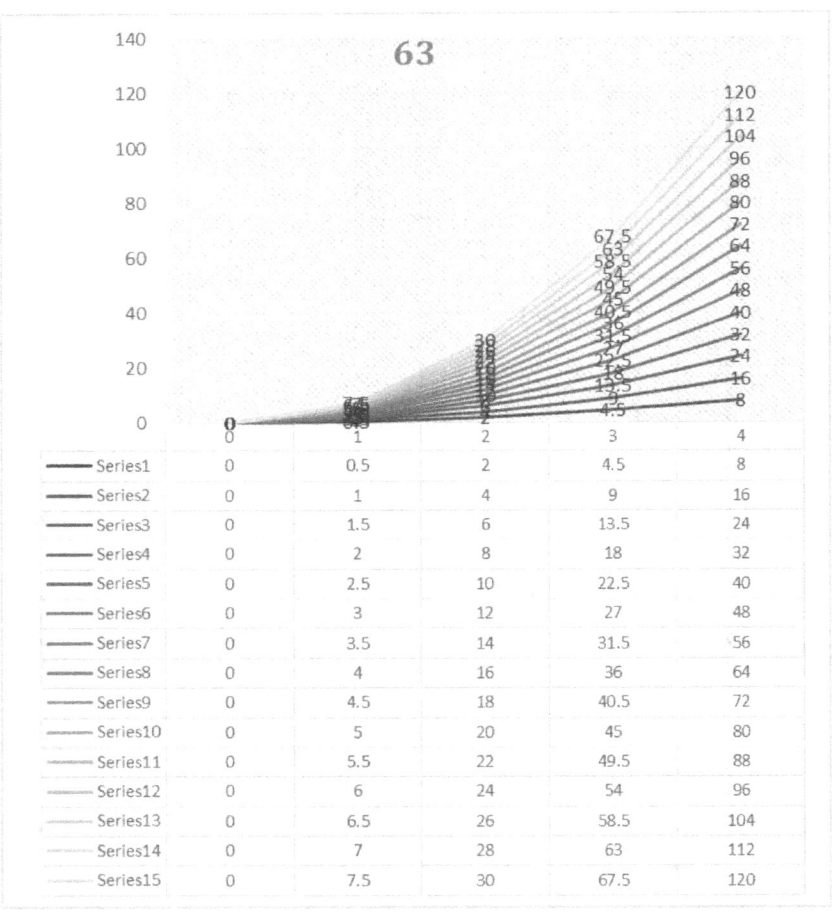

	0	1	2	3	4
Series1	0	0.5	2	4.5	8
Series2	0	1	4	9	16
Series3	0	1.5	6	13.5	24
Series4	0	2	8	18	32
Series5	0	2.5	10	22.5	40
Series6	0	3	12	27	48
Series7	0	3.5	14	31.5	56
Series8	0	4	16	36	64
Series9	0	4.5	18	40.5	72
Series10	0	5	20	45	80
Series11	0	5.5	22	49.5	88
Series12	0	6	24	54	96
Series13	0	6.5	26	58.5	104
Series14	0	7	28	63	112
Series15	0	7.5	30	67.5	120

In Figure 63, a one-dimensional reality consisting of fifteen graphic series is shown. The graphic series shows the acceleration of possible points of the one-dimensional reality. In one-dimensional reality, distances are possible which are in a state of relative rest.

See figure 64.

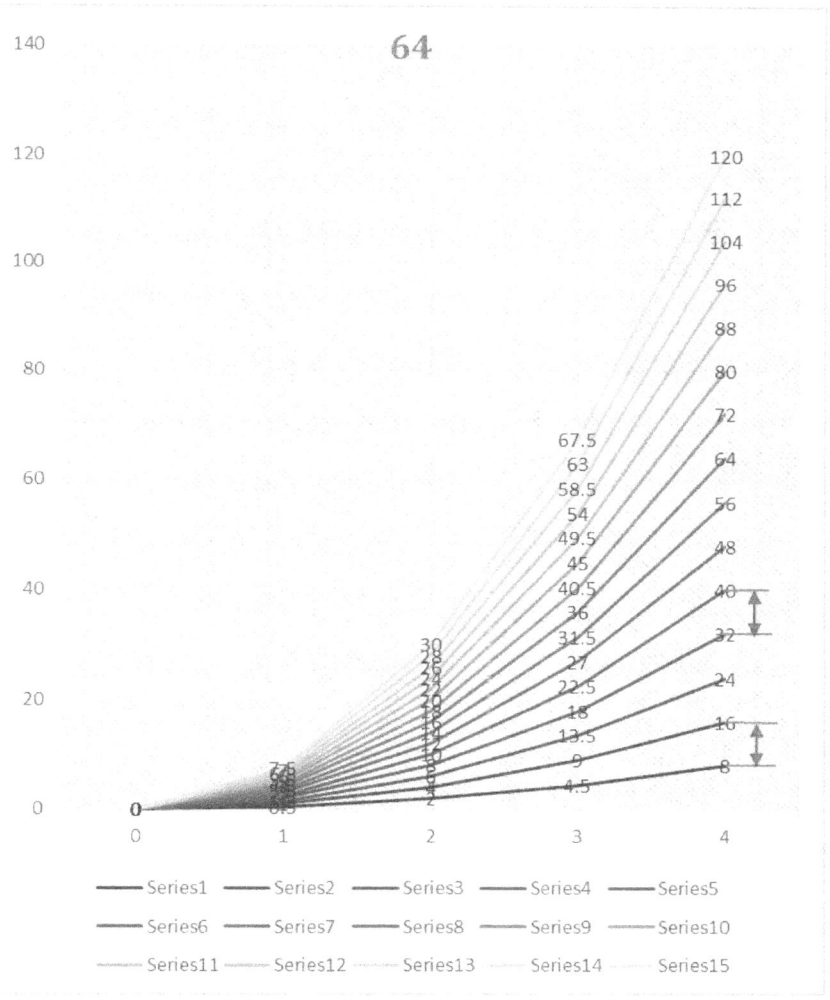

In Figure 64, a one-dimensional reality is shown that has a lifetime of four seconds.

Fifteen graphic series are shown. Bursts start at zero seconds and end at four seconds. The horizontal axis is time, the vertical axis is distance traveled.

Series one is a graph that shows an acceleration of one meter per second squared.

Series two is a graph showing an acceleration of two meters per second squared.

Series three shows an acceleration of three meters per second squared.

For each subsequent series, up the vertical axis, the acceleration is one meter greater.

Series fifteen is at the top, and the acceleration is equal to fifteen meters per second squared.

The vertical distance between the series is always equal to one meter. The meter is a standard, but at the end of each subsequent second, it has different numerical values.

At the end of the fourth second, the numerical value of the distance between the series is equal to the number eight.

Look at the graph, the red arrow and the thin blue lines. The numbers are sixteen and eight. The difference between them is eight.

This eight is a reference distance of one meter, and is present between all series, along the vertical of the fourth second. At the end of the fourth second, the difference between adjacent vertical digits is always the number eight.

At the end of the third second, the difference between the digits that are above each other, vertically, is always equal to the number four and a half. At the end of the third second, the number four and a half, is a standard for a distance equal to one meter.

At the end of the second second, the number two is a standard for a distance equal to one meter.

In one-dimensional reality, physical bodies that exist in a state of rest relative to themselves are possible.

See figure 65.

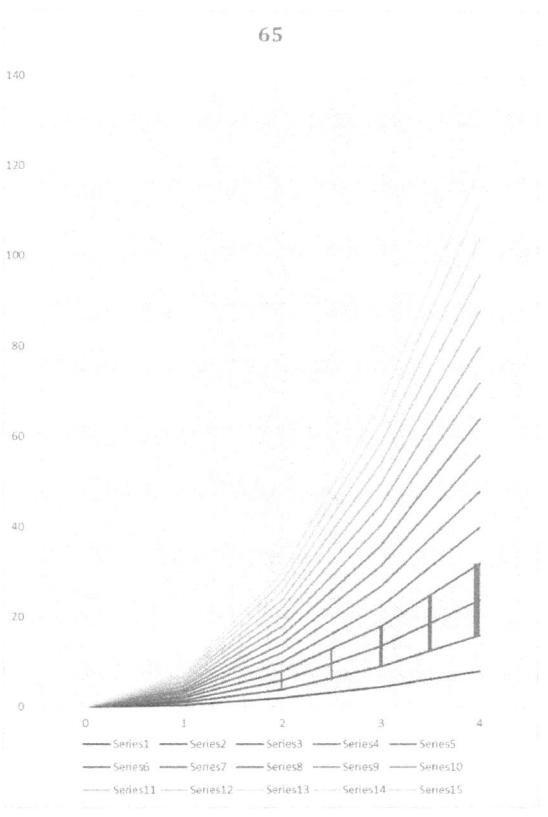

In figure 65, a two meter long body is shown which is at rest relative to itself. The body is shown with a red line.

In one-dimensional reality, physical bodies are possible that exist in a state of rest with respect to themselves, and in a state of rest with respect to other bodies.

See Figure 66.

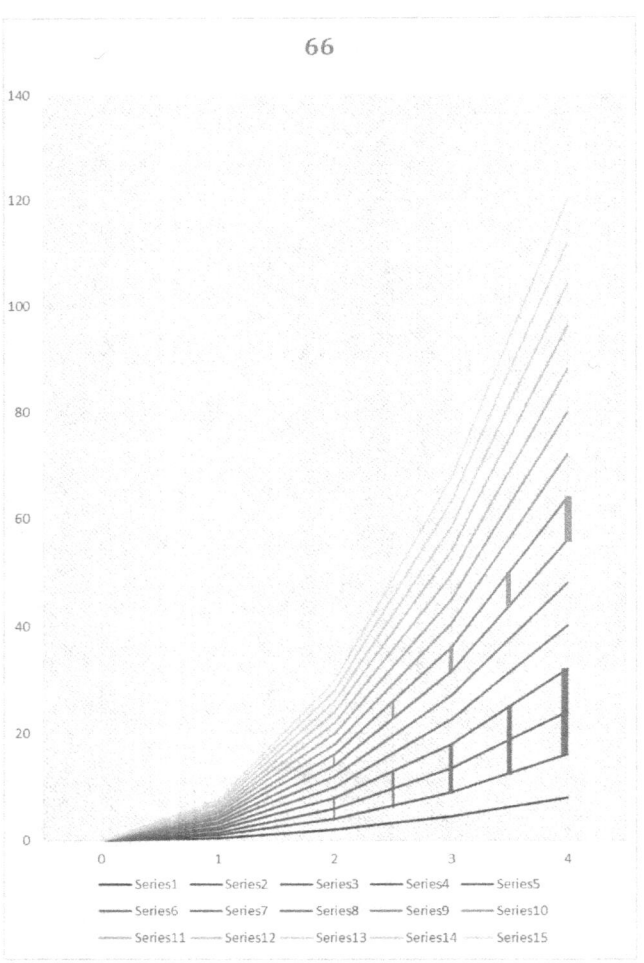

In Figure 66, a one-dimensional reality is shown in which there is one green object and one red object. The red object is two meters long and is located between series two and series four. The green object is one meter long and is located between series seven and series eight. The distance between the red object and the green object is equal to three meters. The green object is at rest relative to itself. The red object is at rest relative to itself. The red object and the green object are at rest relative to each other.

In any one-dimensional reality, uniform rectilinear motion can be performed.

See figure 67.

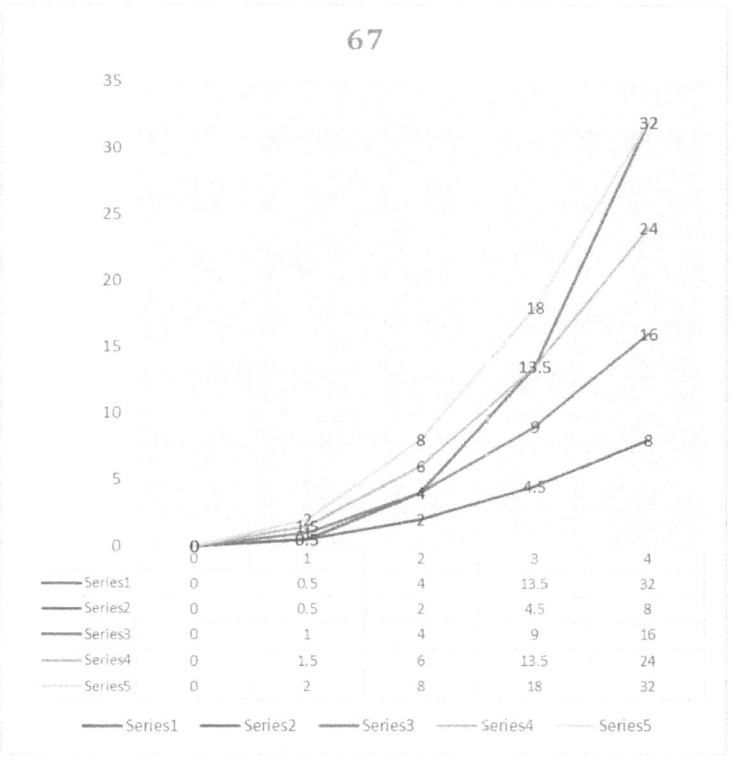

Figure 67 shows uniform rectilinear motion of a red dot, in one-dimensional reality, which has an acceleration coefficient of one meter per second squared. A table with the numerical values of the traveled distance is shown. The red dot moves uniformly in a straight line at a speed of one meter per second.

It is possible to move points that move relative to each other in a uniform straight line.

See figure 68.

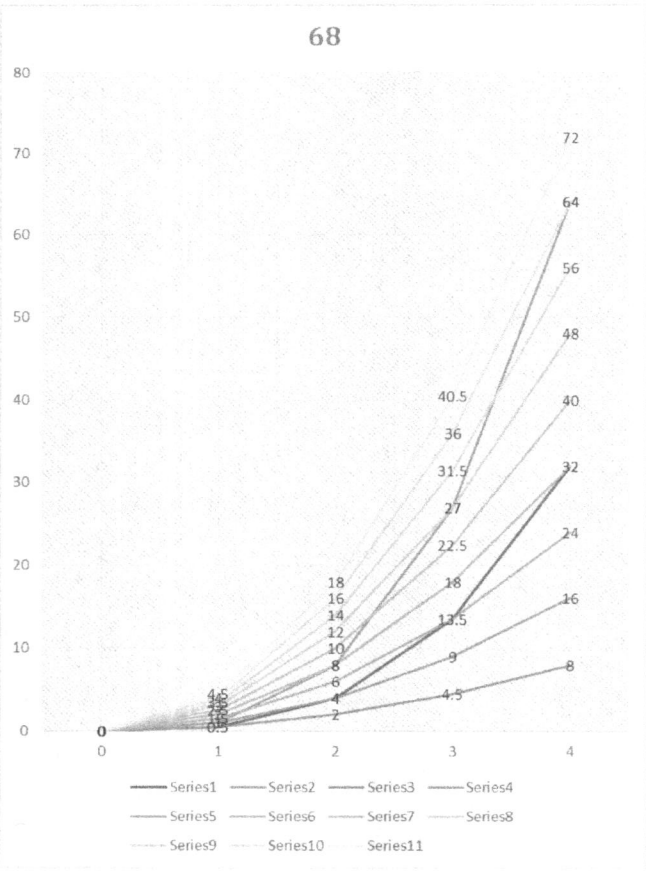

In Figure 68, one-dimensional reality is shown, and uniform rectilinear motion of one red dot and one blue dot.

The red dot moves uniformly in a straight line at a speed of one meter per second, relative to the green one-dimensional reality.

The blue dot moves uniformly in a straight line at a speed of two meters per second relative to the green one-dimensional reality.

The blue dot moves away from the red dot uniformly in a straight line, at a speed of one meter per second.

It is possible to move two or more one-dimensional realities relative to each other.

See figure 69.

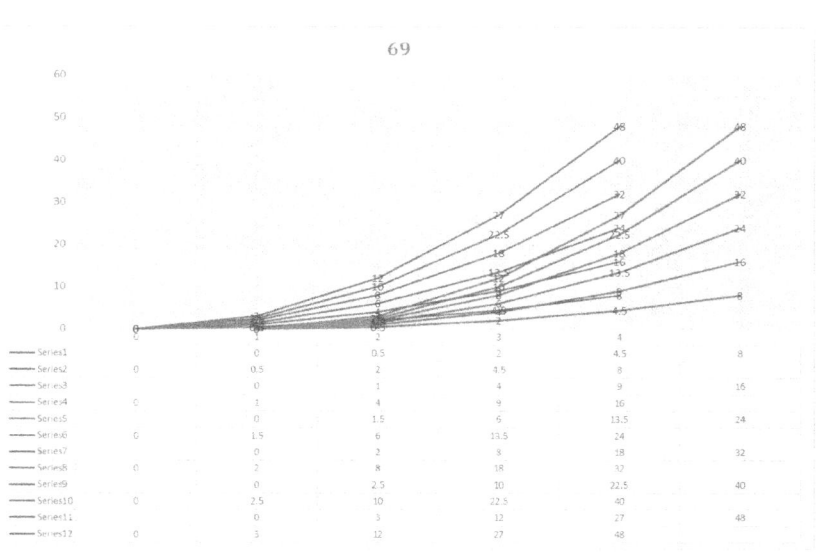

speed of one meter per second.

The red one-dimensional reality exists one second earlier than the blue one.

In a one-dimensional reality, motion with acceleration of any point is possible relative to the whole one-dimensional reality.

See Figure 70.

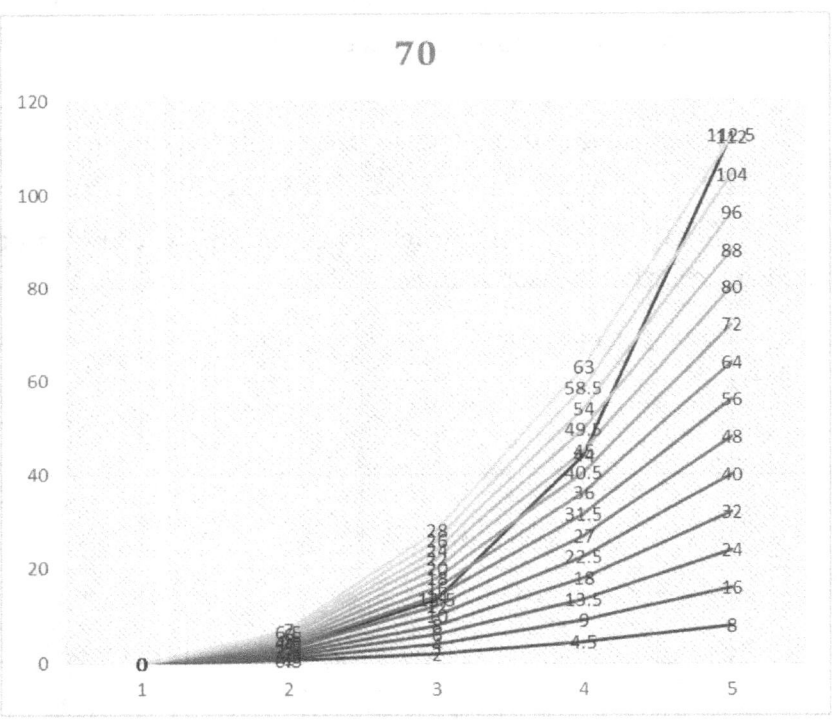

In figure 70, a point is shown that moves with acceleration relative to one-dimensional reality. The point moves in one-dimensional reality with an acceleration of one meter per second squared.

In one-dimensional reality, all different types of motion are possible.

See Fig. 71.

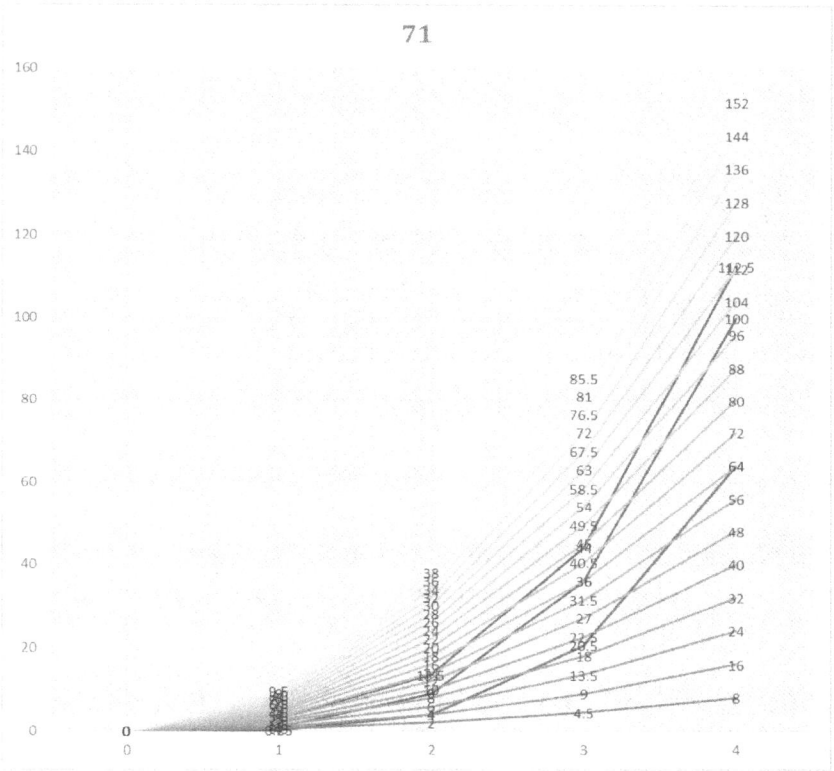

In figure 71, a green one-dimensional reality, two blue dots, and one red dot are shown. The two blues are at rest relative to each other, and are moving with acceleration relative to the green one-dimensional reality. The red dot moves with acceleration relative to the green reality, and it moves uniformly in a straight line relative to the two blue dots.

18. EFFORT. ACCELERATION.

The increase in dimensions of a multidimensional, One Infinite Reality, occurs at an ever- **increasing acceleration** .

Continuously **increasing acceleration** is called **acceleration** .

In the One Infinite Reality there are phenomena that are evidence of the Principle of Sameness.

The first proof is:

The boundaries of the observable universe move away from the center of the observable universe with variable acceleration.

This means that the acceleration of the boundary relative to the center is constantly increasing in a different way. The laws of incremental change are different, and the laws are constantly changing. These are the higher derivatives of the path of time. The amount of higher derivatives is infinitely large.

The center of the observable universe is the planet Earth.

Definition:

The boundary of the observable universe is an infinite number **of places** moving away from planet Earth with an **observable relative velocity** equal to the speed of light.

See Figure 72.

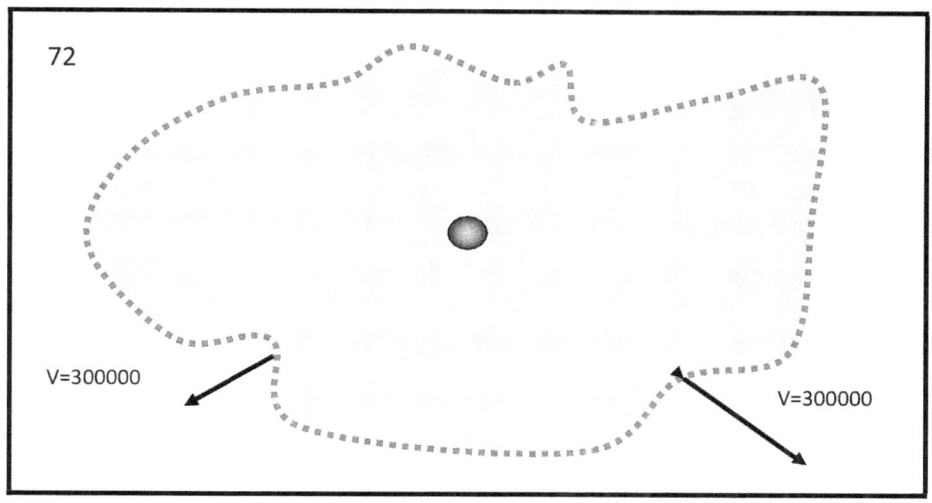

In Figure 72, planet Earth, the observable universe, and the boundaries of the observable universe are shown. Planet Earth is the small sphere in the middle of the figure. Planet Earth is the center of the observable universe. The observable universe is colored light blue. The boundary of the observable universe is shown by the dashed red line. The red line consists of small red squares. The small red squares are **places** in the observable universe. **Places** are **entire parts** belonging to **the entire** observable universe. The concept of **place** replaces the concept of point. I deliberately do not use the term point. The concept of a point is a mathematical abstraction. There are no points in the observable universe. When I use the concept of **place**, I put meaning and content that Newton used in "Mathematical Principles of Physics."

The infinite number **of places** that define the boundaries of the known universe meet a single, necessary and sufficient condition:

They are moving away from the center of the observable Universe with **an observable relative speed**, which is equal to the speed of light, namely, three hundred thousand kilometers per second. The phenomenon **of observable relative velocity** is used only and only as a condition for determining the limit of the "**observable**" Universe. Physical objects moving away at speeds greater than the speed of light cannot be observed using electromagnetic waves that are in the observable optical range of light. The true, absolute motion of the boundary is done with acceleration. In absolute motion with acceleration, there is a moment when the relative observable speed of the physical object, relative to the center, is equal to the speed of light. At this point, this physical object is at the edge of the observable universe. This condition is a tradition in the science of Physics.

The boundary of **the observable** universe is not a sphere. The boundary shown in the figure is not a circle, and is not the true boundary of the observable universe. This is a possible example.

The second proof is:

At different points on the boundary of the observable universe, the acceleration a will be different .

See figure 73.

EINSTEIN'S THIRD MISTAKE

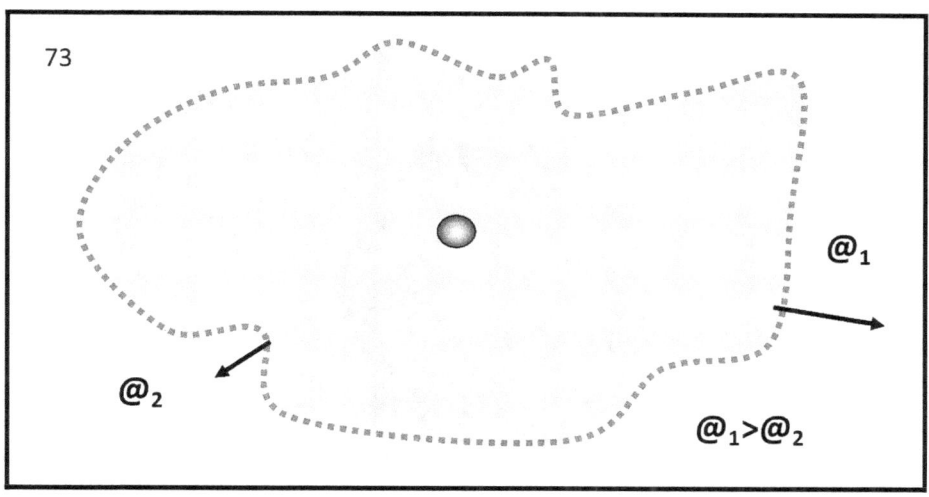

Figure 73 shows different acceleration at the boundary of observable reality. The magnitude of the acceleration is relative to the center of the observable universe. The center of the observable universe is the planet Earth.

The third proof is:

A rod of length equal to the diameter of the planet Earth will accelerate at both ends with an acceleration of nine times eight meters per second squared, relative to its midpoint.

Under this condition, the planet Earth and the rod will be in a state of relative rest.

See figure 74.

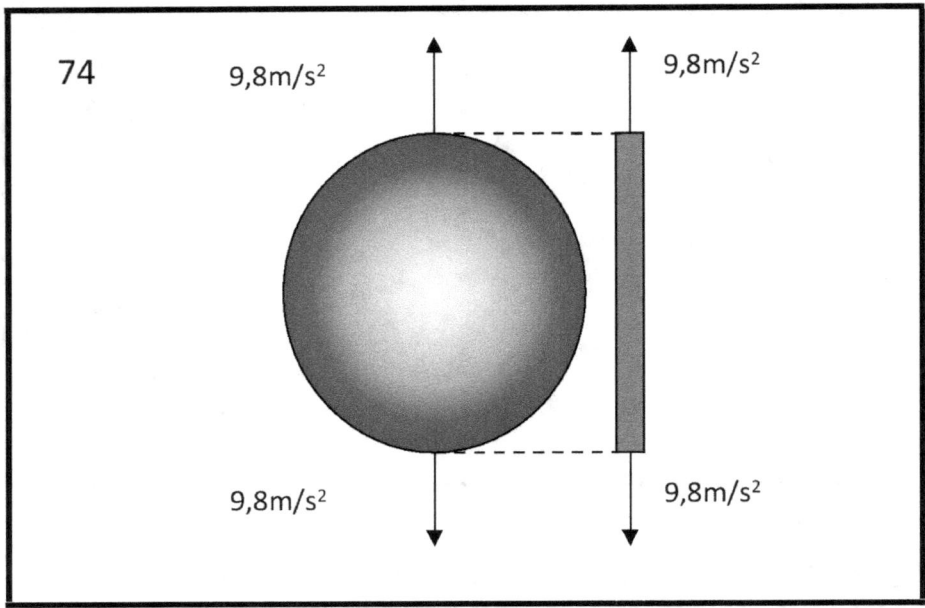

In figure 74, the planet Earth is shown, and a stick. The length of the rod is equal to the length of the diameter of the planet Earth. The two ends of the rod move with root relative to the center of the rod. The acceleration is equal to nine whole eight meters per second squared.

The fourth proof is:

The temperature in the middle of the rod will be higher than the temperature at either end of the rod.

The stick will heat up in the middle.

See Figure 75.

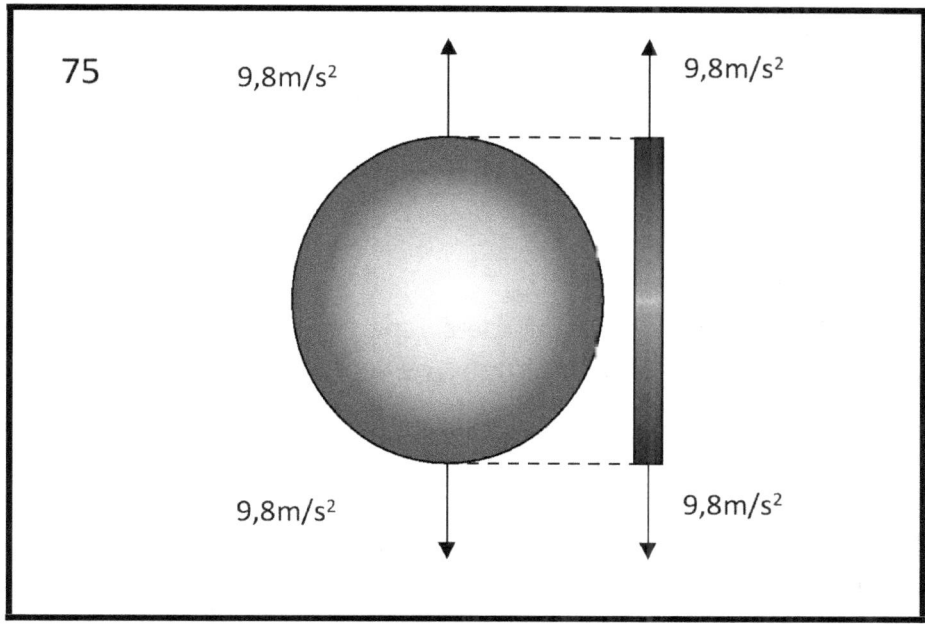

Figure 75 shows the planet Earth and a stick. The length of the rod is equal to the length of the diameter of the planet Earth. The middle of the stick is red because the temperature is high.

19. FIELD OF EFFORT. COMMON FUNDAMENTAL ESSENCE OF THE ONE INFINITE REALITY.

In the fundamental laws of the science of Physics, I define two mutually related quantities, namely – **acceleration** and **effort**.

The acceleration $@$, - is equal to the higher derivatives of the path and time, which are greater than or equal to three.

$$@ = \frac{x}{t^n} \quad \text{......where: } n \geq 3$$

The effort Φ is equal to the product of the mass of the body m and the acceleration $@$.

$$Ф = m.@$$

The letter $Ф$ is from the Slavic-Bulgarian alphabet - Cyrillic.

In **the field of effort the universal interaction between the whole parts of the whole One Infinite Reality** takes place.

It is the only universal connection between the infinite multitude of single whole things which only in this way form the content of the phenomenon of **the entire One Infinite Reality.** The phenomenon of **the whole One Infinite Reality** is possibly reflectable, through and in a state of ever-changing **acceleration**

all One Infinite Reality manifests itself

An ever-changing acceleration, it appears between the discontinuities of **the entire One Infinite Reality** .

A constantly changing acceleration is the cause of the appearance of an infinite **amount** of a particular **quality** , and an infinite **amount** of different **qualities** .

The force is equal to the product of the mass of the whole thing and its acceleration.

$$Ф = m.@$$

Where:

With the letter m we mark the mass of the whole thing.

With the letter $Ф$ from the Slavic-Bulgarian Cyrillic alphabet, we mark **effort**, and with this concept we denote **a fundamental physical quantity** that is equal to the product of the mass of the whole thing and the acceleration.

With the sign $\textcircled{a}$, we mark *acceleration* and with this concept we denote **a fundamental physical quantity** that is equal to or greater than the third derivative of the path from time.

$$\textcircled{a} = \frac{x}{t^n} \ldots\ldots n \geq 3$$

In terms of its historical occurrence, the law of effort, and its relation to acceleration, ranks among the top three laws of classical fundamental physics. Thus, the basic laws of physics are now four.

In terms of its fundamentality and universality, the law of effort encompasses Newton's first three laws.

This gives reason to call it the "zeroth" law of the science of Physics.

The reasons come down to the fact that Newton's laws define a quantitative force interaction between bodies with some specific mass, **whenever , and only when , the force is already manifested and has some specific value** .

In the book "Mathematical Principles of Physics", Newton quite deliberately, regularly uses the terminology "... **action of an applied force** ...".

Newton's deep idea is that this force has appeared and already exists, and can be applied, and acts when applied.

One might argue that Newton's first law does not refer to mutual force interaction. If we carefully analyze the way it is defined, we will come to the conclusion that this is not true.

The law states:

"A body is in a state of rest, or uniform rectilinear motion, when no force is applied to it."

The law can be stated as follows:

"A body is in a state of rest, or uniform rectilinear motion, when it is acted upon by a force equal to zero."

Some reader may object that it makes no sense to speak of a force equal to zero, because it means that no force is applied at all. My answer is that it is possible to apply forces that are equal in magnitude and opposite in direction, and then the result of the action is zero.

Therefore, the inertial movement or the state of relative rest of any particular thing is possible only when the sum of the forces acting on this body is equal to zero.

In other words, from a philosophical standpoint, the concepts of rest and movement denote objective phenomena that are closely related to the result of the action of some specific forces.

It follows that the starting point, or starting position, for determining the phenomenon of rest and the phenomenon of uniform rectilinear motion is **the manifested** force action. It is no accident that Newton used the concept of "action of an applied force."

Newton's second law directly indicates the magnitude of an acting force, expressed as the product of the object's mass and its acceleration.

The law is recorded as follows:

$$F = m.a$$

In Latin, the law reads like this:

„Mutationem motus proportionalem esse vi motrici impressae et fieri secundum lineam rectam qua visilia imprimitur".

From Slavic Bulgarian Cyrillic, via electronic translator:

"The change in the amount of movement is proportional to the applied driving force and is carried out according to the right on which this force acts".

It can be expressed as:

When an m applied driving force acts on a body with mass F, it is in a state of motion with constant acceleration a.

It is not necessary to make an analysis to see that the law indicates the quantity of the force when it **has already manifested itself** and is of some constant concrete value.

Newton's third law written in Latin:

> „Actioni contrariam semper et aequalem esse reactionem: sive corporum duorum actiones in se mutuo semper esse aequales et in partes contrarias dirigi"

From Slavic Bulgarian Cyrillic, via electronic translator:

"The action is always equal and opposite to the counteraction, in other words, the interactions of two bodies, one on the other, between themselves, are equal and directed in opposite directions."

Said in this way, it shows that when a body is *acted* upon by a force, from another body, then the body reacts with a force which is equal in magnitude and opposite in direction.

In this case, we again notice that in Newton's third law it is again a question of a force that has already **manifested itself** and already **operates** with some particular constant magnitude.

We ask only one, but extremely important question:

How does it **appear** ? the action of the force F ?

Our answer, which is a result of the effort field hypothesis created, is:

The amount of interaction between things appears in a field of effort.

See Figure 76.

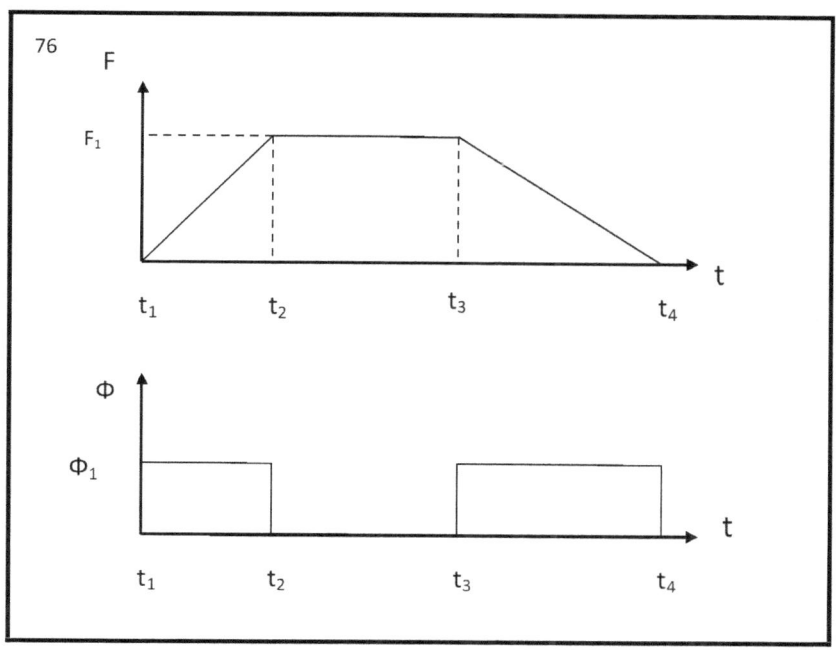

76

In figure 74, it is shown how, in the time interval $t_2 - t_1$, the force appears F, and how it increases from zero to some value F_1, see the above coordinate system.

In the same time interval $t_2 - t_1$, the

phenomenon of constant acting force is observed Φ_1, which is shown on the bottom coordinate system.

In the time interval $t_4 - t_3$, the force decreases from some value F_1, to zero (upper graph) and again appears as a constant acting force of magnitude Φ_1, which is shown on the second (lower) coordinate system.

Once again, we must emphasize that the considerations expressed in this way give us a reason to declare the law of effort $\Phi = m.@$ as the "zero" law of Physics, which precedes Newton's laws.

As a law that operates in the absolute foundation of **all One Infinite Reality**.

As a law that is the reason for the appearance of Newton's first three laws.

As a law that defines the phenomenon **field of effort**.

As a law that opens the door behind which the creation of a general field theory is possible.

This law is essentially an introduction to GENERAL FIELD

THEORY.

The term "**field of effort**" serves to denote a phenomenon existing throughout **the One Infinite Reality,** the essence of which has a universal fundamental character.

It is possible that this fundamental, as yet physically unexplained and unclear field, may turn out to be the basis and key to the deep secrets of the Absolute Movement and its appearing entities in the direction of Space, Time and the way in which they are constructed and exist in the real things of Nature.

In purely practical terms, technological mastery of **the field of effort** would provide humanity with limitless informational freedom to communicate with **all of One Infinite Reality** and its constituent **parts** absolutely simultaneously.

If, however, this task of technological mastery of distant action turns out to be the most unattainable dream, then humanity will forever remain captive to the limitations imposed on it by Time, Space and Movement.

Optimism inspires the modern development of the philosophical-physical conception of reality, which gives hope that this will not happen.

These two new quantities - **effort and acceleration**, and the relationship between them allow us to renew the content of some fundamental categories of physics.

For example:

Force, defined by Newton's second law F, has a regular relationship with relative interaction and its quantitative essence.

The effort $Ф$, expresses the amount of absolute interaction.

Heavy mass – the amount of breaks in the continuum.

The inertial mass – the continuity of storage of the link between breaks.

However, these questions, as well as some higher derivatives of the time path, should be the subject of a separate scientific analysis.

20. NEWTON, GRAVITY AND FIELD OF EFFORT.

The principle of uniformity shows that a force of gravitational attraction, as represented by Newton, does not exist. What Newton called the force of gravitational attraction is motion with acceleration. The Sun and the planets of the solar system increase their radii at different rates. The increase of the radii with different acceleration is done relative to the center of the particular planet and the center of the Sun.

The solar system increases its radius with acceleration. The acceleration of the periphery of the solar system is relative to the center of the solar system. The center of the solar system coincides with the center of the Sun.

Newton's law of gravitational attraction holds true within the confines of the solar system. But what Newton called gravitational attraction is a movement of pushing, pushing, with acceleration.

The pushing motion, pushing with acceleration, occurs and takes place in the field of effort. Acceleration occurs, which is the reason for the appearance of a pushing force. The magnitude of the pushing force within the limits of the solar system is calculated by the law of gravitational attraction stated by Newton. Elsewhere in the One Infinite Reality, the magnitude of the repulsive force will be different from the repulsive force that operates within the confines of the solar system. This means that Newton's law of gravity will be different.

The amount of "Newton's other laws" in the One Infinite Reality is infinitely great.

The pushing force appears in the field of effort and depends on the law according to which the acceleration changes.

In the One Infinite Reality, the number of possible laws by which acceleration is changed is infinitely large.

21 TIME

In the One Infinite Reality exists the Phenomenon of Time. The essence of the Time Phenomenon is movement with increasing acceleration.

A fundamental property of the Time Phenomenon is integral irreversibility.

www.ingramcontent.com/pod-product-compliance
Lightning Source LLC
Chambersburg PA
CBHW050003230526
45465CB00003BB/1232